5

SERIES IN THEORETICAL AND APPLIED MECHANICS

Edited by R.K.T. Hsieh

SERIES IN THEORETICAL AND APPLIED MECHANICS
Editor: R. K. T. Hsieh

Volume 1 Nonlinear Electromechanical Effects and Applications
by G. A. Maugin

Volume 2 Lattice Dynamical Foundations of Continuum Theories
by A. Askar

Volume 3 Heat and Mass Transfer in MHD Flows
by E. Blūms, Yu. Mikhailov, R. Ozols

Volume 4 Mechanics of Continuous Media (in 2 volumes)
by L. I. Sedov

Volume 5 Inelastic Mesomechanics
by V. Kafka

Inelastic Mesomechanics

Vratislav Kafka

Author

V. Kafka

Institute of Theoretical and Applied Mechanics of the Czechoslovak Academy of Sciences Vyšehradská 49, 12849 Praha 2, Czechoslovakia

Series Editor-in-Chief

R. K. T. Hsieh

Department of Mechanics, Royal Institute of Technology S-10044 Stockholm, Sweden

Published by

World Scientific Publishing Co. Pte. Ltd.
P. O. Box 128, Farrer Road, Singapore 9128
U. S. A. office: World Scientific Publishing Co., Inc.
687 Hartwell Street, Teaneck NJ 07666, USA

Distributed by

John Wiley & Sons Ltd. (in Europe)
D A Book Pty Ltd. (in Australia)
Allied Publishers Pvt Ltd. (in India)

Library of Congress Cataloging-in-Publication Data is available.

INELASTIC MESOMECHANICS

ISSN 0218-0235
ISBN 9971-50-505-3

CONTENTS

Inelastic Mesomechanics

INTRODUCTION

Most materials with which technical mechanics has to deal are heteregeneous materials that seem to be homogeneous on the macro-scale and are composed of several material constituents. If we are interested only in quasistatic elastic deformation processes, it does not seem to be of a high practical interest to analyse the heterogeneity of the material - it is sufficient to measure the elastic constants. On the other hand, in problems of strength, dynamics, temperature changes and inelasticity, it is more difficult to formulate a satisfactory mathematical model, and a better understanding and description of the macroscopic mechanical properties calls for an analysis based on the knowledge of the heterogeneous composition.

Generally speaking the theoretical approaches, from which the mechanical problem of heterogeneous, macroscopically homogeneous materials has been attacked, can be grouped as follows:

a/ Phenomenological Models

In this case the questions of heterogeneity are not at all discussed and the material under study is described as homogeneous. This group covers elasticity, plasticity, rheology and more and more general variants of continuum mechanics as they are surveyed e.g. in Eringen /2/.

b/ Theory of Internal Variables

This promissing approach describes complicated material properties by adding the so called "internal variables" to the common variables used in classical continuum mechanics. The generality of the resulting relations can be reduced by the application of the laws of thermodynamics. Even without specification of the physical meaning of the internal variables some interesting results can be achieved - - cf. Coleman and Gurtin 8/, Kratochvíl and Dillon 42/, Bruhns and Müller 6/. A specific variant of this approach is the so called "Endochronic Theory", where a special - intrinsic - kind of time is introduced to describe the changes of the instantaneous material properties - cf. Valanis 71, 72/.

c/ Micromechanical Models

The above title is usually used for such theories, which are based on the analysis of the structure of the material in question. A number of results in this line can be found e.g. in Kunin 44,45/ Sendeckyj 50/. This group of models can be divided in several subgroups:

Deterministic Models use for their basis the notion of a "cell" or a "representative volume element" with uniquely defined geometry of composition, which is repeated throughout the body. Very often this geometry means simplification of reality. This approach is suitable especially for materials with simple structure. Some fundamental results can be found in Hill 18/, Hashin 14/, Vanin 73/, Christensen 7/.

Statistical Models admit the geometry of composition to be arbitrarily complicated and the microstructure is described by statistical momenta. The limiting factors here are the input information which is difficult to acquire and the complicated system of equations that are arrived at in case of inelasticity. Such models can be found e.g. in Beran 3/, Axelrad 1/.

Physical Models are based on the very microscopic scale of description, on the scale of atomic lattice and its dislocations. The high regularity of the lattice makes the analysis relatively easy and this is probably one of the reasons, why this approach is so much elaborated. The problem is that between this scale and the macroscopic scale, there is a number of scales of irregularities, which influence the macroscopic behaviour substantially. In spite of it some qualitative conclusions can be reached in this way even for macroscopic properties.

The presented model is aimed at quasistatic problems of inelasticity and it starts from a philosophy that is based on the following points of view:

Mechanical properties of a heterogeneous material can be observed and described on different scales that can be named: macroscale, mesoscale, microscale, atomic scale etc. We believe that for the understanding of the properties on the macroscale the most reasonable and fruitful approach is an analysis that is based on the one-step-down scale. i.e. on the mesoscale. Hence the title of the book. However. we have also recognized that it is not acceptable to work only with the averages of stress and strain on the mesoscale, the influence of the still lower scales must be included in terms of fluctuations. Otherwise the results lead to contradictions with experimental evidence.

The problem of any micromechanical or mesomechanical approach is that the basis of the analysis - the mechanical properties of the material constituents and the geometry of their composition - are in fact difficult to determine. The geometry is very often complicated, the in-situ properties of the material constituents differ in many cases substantially from those of a homogeneous sample, the interfaces also have a strong influence. We believe therefore that the only way out is to get as much information as possible from macroscopic experiments, i.e. from a solution of the identification problem that is based on a mathematical analysis of the stress-strain curves or the flow-curves observed on macroscopic samples. Considering this input information as the basic information seems to be as natural as measuring Young's modulus instead of calculating it from the microstructural properties.

The possibility of getting easily the necessary input information, related really and directly to the heterogeneous material in question, is looked upon as a fundamental requirement on the formulation and construction of our mathematical models. The considerations that follow respect this point of view.

In the three fundamental parts of the book the mesomechanical aspect is applied in three different ways:

In the first part, called "Mesomechanics of Inelastic Deformations" the volume of the heterogeneous material is divided into the subvolumes of the individual material constituents and the evolution of stress-fields and strain-fields is the main topic of the analysis.

In the second part, called "Mesomechanical Limit Analysis" the volume is again divided into the subvolumes of the individual material constituents, but the topic of the analysis is not the evolution of the fields, but their limit states that cannot be overpassed without rupture.

The third part, called "A Non-Local Criterion of Strength" is aimed at such processes where the redistribution of internal stresses is not substantial. This is the case e.g. in brittle fracture. Here the body is not divided into the subvolumes of the material constituents, but for any macroscopic point its specific spherical neighbourhood with the stress-field contained in it is the basis for analysis. Hence the non-local character of the criterion. The important influence of the surface layer is taken into account by attributing to its points other values of the criterion parameters than to the interior points.

SYMBOLIC NOTATION

$\mathbf{X}$	macroscopic coordinate
$\mathbf{x}$	microscopic coordinate
t	time
$N(\mathbf{X})$	small spherical macroscopic neighbourhood of $\mathbf{X}$
$\bar{\sigma}_{ij}(\bar{\varepsilon}_{ij})$	macroscopic stress (strain) tensor /averages in $N(\mathbf{X})$ — functions of $\mathbf{X}$ /
$\sigma_{ijn}(\varepsilon_{ijn})$	mesoscopic stress (strain) tensor /averages in the n-th material constituent comprised in $N(\mathbf{X})$ - functions of $\mathbf{X}$ /
$\sigma'_{ij}(\varepsilon_{ij})$	microscopic stress (strain) tensors /functions of $\mathbf{x}$ /
$\sigma^*_{ijn}(\varepsilon^*_{ijn})$	stress (strain) tensor that corresponds in the n-th material constituent to the strain (stress) history identical with macroscopic strain (stress) history
σ^r_{ij}	residual stress
$i, j, k, l\ (\alpha, \beta)$	indices the repetition of which means (does not mean) summation
$\sigma = \frac{1}{3}\sigma_{ii}$ $(\varepsilon = \frac{1}{3}\varepsilon_{ii})$	isotropic part of the stress (strain) tensor
$s_{ij} = \sigma_{ij} - \delta_{ij}\sigma$ $(e_{ij} = \varepsilon_{ij} - \delta_{ij}\varepsilon)$	deviatoric part of the stress (strain) tensor
δ_{ij}	Kronecker's delta
v_n	volume fraction of the n-th material constituent
$\varepsilon'_{ijn} = \varepsilon_{ijn} - \bar{\varepsilon}_{ij}$	in the A-model
$\varepsilon'_{ijn} = \varepsilon_{ijn} - \varepsilon^*_{ijn}$	in the B-model
$\sigma'_{ijn} = \sigma_{ijn} - \sigma^*_{ijn}$	in the A-model
$\sigma'_{ijn} = \sigma_{ijn} - \bar{\sigma}_{ij}$	in the B-model
$f^e/f^\varepsilon\,(f^s/f^\sigma)$ $[\varphi^e/\varphi^\varepsilon\,(\varphi^s/\varphi^\sigma)]$	distribution functions for the deviatoric/isotropic parts of the strain (stress) tensor in the A-model [B-model] of the material with macroscopic isotropy
$\eta_n/\eta^o_n\,(\chi_n/\chi^o_n)$	structural parameters for the deviatoric/isotropic parts in the A-model (B-model) of the material with macroscopic isotropy
$E_n\ (\bar{E})$	Young's modulus of the n-th material constituent (of the composite)

$\nu_n\ (\bar{\nu})$ Poisson's ratio of the n-th material constituent (of the composite)

$\mu_n = (1+\nu_n)/E_n$ /analogously for $\bar{\mu}, \bar{\varrho}, \bar{\nu}, \bar{E}$ /

$\varrho_n = (1-2\nu_n)/E_n$

$D_n\ (I_n)$ constant or variable coefficients in the constitutive equations for the deviatoric (isotropic) tensor parts of the n-th material constituent

T temperature

τ_n some specific parameter in the constitutive equation of the n-th material constituent

$\alpha_n (\bar{\alpha})$ coefficient of thermal expansion of the n-th material constituent (of the composite)

$\dot{h}$ defined by eq. (53)

H coefficient of deviatoric viscosity

λ scalar measure of plastic deformation

β coefficient of the Bingham - type deformation

$k = c\sqrt{3}/2$ plastic limit /cf. eq. (103) /

Symbols with local meaning are defined in the text.

I. MESOMECHANICS OF INELASTIC DEFORMATION

The type of material that we are going to analyse is supposed to be homogeneous on the macroscale and heterogeneous on the meso-scale. It is composed of several material constituents that are defined by their specific material properties and by the geometry of their infrastructure that is statistically regular. This means that the infrastructure can be different in different macroscopic volumes, but the respective statistical description is the same up to the degree of accuracy that is important for macroscopic response. The infrastructures of all the material constituents form the structure of the material. The sum of the subvolumes of the material constituents equals the volume of the body and the volume fraction of any material constituents is the same in any macroscopic volume. By one mathematical model only such process can be described, for which - to a sufficient degree of accuracy - the structure remains statistically stable, i.e. the volume fractions and the statistical description of the infrastructures remain unchanged. Substantial changes in the structure require determination of new parameters or even some new type of the mathematical model.

In our analysis we use three scales for the description of stress- and strain-fields:

The first scale is the macroscopic one with the macroscopic stress $\bar{\sigma}_{ij}(\boldsymbol{X},t)$ and strain $\bar{\varepsilon}_{ij}(\boldsymbol{X},t)$, which depend on the macroscopic coordinate $\boldsymbol{X}$ and time t. It is commonly used and does not need any explanation. /In our approach time t is to be understood as the measure of progress of the deformation process. In the case of viscosity it is suitable to identify t with the current time. In some other cases some special "intrinsic" time can be preferable/.

Further, we introduce the notion of a small spherical macroscopic neigbourhood $N(\boldsymbol{X})$ of the macroscopic point $\boldsymbol{X}$. $N(\boldsymbol{X})$ is small in regard to the body and the space - variations of $\bar{\sigma}_{ij}$ and $\bar{\varepsilon}_{ij}$, but it is macroscopic, i.e. its diameter is at least one order higher than the characteristic length of the structure.
For the description of the internal phenomena in $N(\boldsymbol{X})$ we use microscopic coordinates $\boldsymbol{x}$. We can use them for the description of the infrastructures of the material constituents in $N(\boldsymbol{X})$ and for the evaluation of the average values of stress and strain in individual material constituents. These average values are different for different material constituents, but for one material constituent comprised in one $N(\boldsymbol{X})$ they are unique. Therefore, they are functions of $\boldsymbol{X}$, but not of $\boldsymbol{x}$. For the n- material constituent they are described as $\sigma_{ijn}(\boldsymbol{X},t)$, $\varepsilon_{ijn}(\boldsymbol{X},t)$. This is the second scale,

called mesoscale.

However, the stress-field and strain-field in the n-th material constituent in one $N(\mathbf{X})$ need not be homogeneous and generally they vary with $\mathbf{x}$. Hence, they must be described as $\sigma_{ij}(\mathbf{X}, \mathbf{x}, t)_n$, $\varepsilon_{ij}(\mathbf{X}, \mathbf{x}, t)_n$. This is the third scale, called microscale.

From the above definitions it follows:

$$\langle \sigma_{ij}(\mathbf{X}, \mathbf{x}, t)_n \rangle_n = \sigma_{ijn}(\mathbf{X}, t)$$

$$\langle \varepsilon_{ij}(\mathbf{X}, \mathbf{x}, t)_n \rangle_n = \varepsilon_{ijn}(\mathbf{X}, t)$$

where $\langle \cdots \rangle_n$ means the volume average in the n-th material constituent.

The mesoscopic variables are basic for the analysis, the influence of the microscopic variables will be expressed in terms of fluctuations and the relations between the macroscopic variables /i.e. the macroscopic constitutive equations/ represent the result of the analysis.

The conditions of equilibrium and of continuity of displacements lead to the well known equations:

$$\sum_{n=1}^{N} v_n \sigma_{ijn} = \bar{\sigma}_{ij} \qquad (1)$$

$$\sum_{n=1}^{N} v_n \varepsilon_{ijn} = \bar{\varepsilon}_{ij} \qquad (2)$$

where v_n is the volume fraction of the n- material constituent and

$$\sum_{n=1}^{N} v_n = 1 \qquad (3)$$

Furthermore, the following equation holds true:

$$\overline{\sigma_{ij} \varepsilon_{ij}} = \bar{\sigma}_{ij} \bar{\varepsilon}_{ij} \qquad (4)$$

For the first time the last equation was deduced by R. Hill /17/ and later on discussed in Hill /18, 19/, Kafka /25, 35/ and Majumdar and Laughlin /48/.

In what follows we accept the fundamental theorem of the theory of internal variables that any current state of the material can be described by a finite number of variables. Hence, we can consider

$$\pi(S_p) = \bar{\sigma}_{ij} \bar{\varepsilon}_{ij} - \overline{\sigma_{ij} \varepsilon_{ij}} = 0 \quad (p = 1, ---, P) \qquad (5)$$

as a function of variables S_p that include the internal variables. This function equals zero throughout any process. P is some finite number of variables.

Consequently any variation of $\pi(S_p)$ equals zero:

$$\delta\, \pi(S_p) = \sum_{p=1}^{P} \frac{\partial \pi}{\partial S_p} \delta S_p = 0 \qquad (6)$$

Furthermore, let us suppose that we use all the constraints that are valid among variables S_p and that we eliminate some of them and restrict their number to the independent variables only. In this way function $\pi(S_p)$ will change into another form that we express as

$$\pi'(S_p) \qquad (p = 1, ---, Q)\ ,$$

where $Q(\leq P)$ is the number of the independent variables.

Then it holds true for any variation of $\pi'(S_p)$:

$$\delta\, \pi'(S_p) = \sum_{q=1}^{Q} \frac{\partial \pi'}{\partial S_q} \delta S_q = 0 \qquad (7)$$

for any set of independent variations δS_q and therefore

$$\frac{\partial \pi'}{\partial S_q} = 0 \qquad (8)$$

The number of the constraints was $P-Q$ and with Q equations (8) the total number of equations is P , equalling the number of the unknowns S_p .

I. 1 Materials With Macroscopic Isotropy in the Virgin State

I. 1. 1 General Description of Strain- and Stress - Distribution

Without any loss of generality the distribution of strain and stress can be described alternatively in the following two dual ways:

A/ $$e_{\alpha\beta}(X, x, t)_n = \bar{e}_{\alpha\beta}(X, t) + f^{e}_{\alpha\beta}(X, x, t)_n \; e'_{\alpha\beta n}(X, t) \qquad (9)$$

$$\varepsilon\, (X, x, t)_n = \bar{\varepsilon}\, (X, t) + f^{\varepsilon}\, (X, x, t)_n \; \varepsilon'_n\, (X, t)$$

$$s_{\alpha\beta}(X, x, t)_n = s^{*}_{\alpha\beta n}(X, t) + f^{s}_{\alpha\beta}\, (X, x, t)_n \; s'_{\alpha\beta n}(X, t)$$

$$\sigma\, (X, x, t)_n = \sigma^{*}_n\, (X, t) + f^{\sigma}\, (X, x, t)_n \; \sigma'_n\, (X, t)$$

or:

$$\text{B/} \quad s_{\alpha\beta}(X,\varkappa,t)_n = \bar{s}_{\alpha\beta}(X,t) + \varphi^s_{\alpha\beta}(X,\varkappa,t)_n\, s'_{\alpha\beta n}(X;t) \tag{10}$$

$$\sigma(X,\varkappa,t)_n = \bar{\sigma}(X,t) + \varphi^\sigma(X,\varkappa,t)_n\, \sigma'_n(X,t)$$

$$e_{\alpha\beta}(X,\varkappa,t)_n = e^*_{\alpha\beta n}(X,t) + \varphi^e_{\alpha\beta}(X,\varkappa,t)_n\, e'_{\alpha\beta n}(X,t)$$

$$\varepsilon(X,\varkappa,t)_n = \varepsilon^*_n(X,t) + \varphi^c(X,\varkappa,t)_n\, \varepsilon'_n(X,t)$$

where $e_{\alpha\beta}/\varepsilon\,(s_{\alpha\beta}/\sigma)$ is the deviatoric/isotropic part of the strain (stress) tensor; $\bar{e}_{\alpha\beta}/\bar{\varepsilon}\,(\bar{s}_{\alpha\beta}/\bar{\sigma})$ is the macroscopic deviatoric/isotropic part of the strain (stress) tensor; $s^*_{\alpha\beta n}/\sigma^*_n\,(e^*_{\alpha\beta n}/\varepsilon^*_n)$ is the deviatoric/isotropic part of the stress (strain) tensor that corresponds in the n- material constituent to the strain (stress) history identical with the macroscopic strain (stress) history; the distribution functions f and φ are the only expressions on the right-hand sides of the above equations that depend on $\varkappa$ and by definition they have unit norm, i.e. for any of the four expressions f or φ it holds:

$$\langle f \rangle_n = 1 \tag{11}$$

$$\langle \varphi \rangle_n = 1$$

There is no summation with the repeated indices α, β.

The meaning of the remaining variables turns out after performing the averaging procedure $\langle - - \rangle_n$ in all of the equations:

$$\text{A/} \quad e'_{\alpha\beta n} = e_{\alpha\beta n} - \bar{e}_{\alpha\beta} \tag{12}$$

$$\varepsilon'_n = \varepsilon_n - \bar{\varepsilon}$$

$$s'_{\alpha\beta n} = s_{\alpha\beta n} - s^*_{\alpha\beta n}$$

$$\sigma'_n = \sigma_n - \sigma^*_n$$

$$\text{B/} \quad s'_{\alpha\beta n} = s_{\alpha\beta n} - \bar{s}_{\alpha\beta} \tag{13}$$

$$\sigma'_n = \sigma_n - \bar{\sigma}$$

$$e'_{\alpha\beta n} = e_{\alpha\beta n} - e^*_{\alpha\beta n}$$

$$\varepsilon'_n = \varepsilon_n - \varepsilon^*_n$$

I.1.2 The Fundamental Theorem of the Mathematical Model

The equations of the preceding paragraph are quite general, descriptive of any arbitrarily complicated reality. It is necessary to simplify them to create an operative mathematical model. The simplifying theorem that forms our concept is that only such distribution functions f and φ are taken into account, which are independent of t. There are two possible interpretations of this step:

a/ The second addends on the right-hand sides of eqs.(9) and (10) represent fluctuations that - for a fixed X - depend on x and t. We restrict the analysis to such cases only, where the fluctuations are described in terms of products of two functions, one depending on x, the other on t.

b/ The distribution functions f and φ can be understood as hypersurfaces of the relative distribution of fluctuations. A necessary condition for these hypersurfaces to be time--independent is the time-independence of structure of the material. We consider time-independent structures only and restrict our description only to such models, where the condition of time-independence of structure is not only necessary, but also sufficient for the time-independence of the hypersurfaces of relative distribution of fluctuations. This is a restriction to generality, but still, the freedom that remains is high: the macroscopic values, the volume averages in the individual material constituents and the extent of fluctuations are free to change.

More important than the interpretations of the theorem are the properties of the resulting mathematical model. As will be shown later on, they are:

- The model covers as special cases all possible combinations of compact and loose infrastructures of the material constituents. /To be compact means such property of the infrastructure that rigidity of the respective material constituent leads to rigidity of the composite. An infrastructure that is not compact is called loose/. Furthermore, it spans the interval between the natural limits resulting from the hypothesis of homogeneous stress /Reuss's solution/ and that of homogeneous strain /Voigt's solution/. Both the last named models result as limiting special cases of our general model.

- It takes into account not only the different average values of stress and strain in individual material constituents /mesoscale/, but also the influence of the lower scales in terms of fluctuations. Only in the case of inclusions in a compact matrix the description of stress and strain in the inclusions turns out as homogeneous in one $N(X)$.

- It admits arbitrarily complicated structures and a broad variety of constitutive equations of the material constituents.

- It enables relatively easy solutions to the identification problems based on simple macroscopic experiments.

- It leads to results that are compatible exactly or very closely with a number of other mathematical solutions based on quite different approaches.

- A number of applications can be shown to agree with experimental evidence.

- The model is the simplest one with the above specified properties.

I.1.3 Model Description of Strain- and Stress- Distribution

Using the fundamental theorem specified in the preceding paragraph we can simplify the description substantially. All the functions f and φ are now independent of t, i.e. they do not change in the course of a deformation process, and from the a priori assumption that the material is macroscopically homogeneous in the virgin state, they are now independent of X as well. Furthermore, in this section we analyse only materials with macroscopic isotropy, i.e. such materials, in which any $N(X)$ has its statistical properties unchanged after rotation. Therefore, all the functions f and φ do not change with orthogonal transformation of axes, which means that they are scalars. Hence, it holds:

A/
$$
\begin{aligned}
e_{\alpha\beta}(X,x,t)_n &= \bar{e}_{\alpha\beta}(X,t) + f^e(x)_n \, e'_{\alpha\beta n}(X,t) \\
\varepsilon(X,x,t)_n &= \bar{\varepsilon}(X,t) + f^{\varepsilon}(x)_n \, \varepsilon'_n(X,t) \\
s_{\alpha\beta}(X,x,t)_n &= s^*_{\alpha\beta n}(X,t) + f^s(x)_n \, s'_{\alpha\beta n}(X,t) \\
\sigma(X,x,t)_n &= \sigma^*_n(X,t) + f^{\sigma}(x)_n \, \sigma'_n(X,t)
\end{aligned}
\qquad (14)
$$

and quite similarly for the B- model, in eq. (10) there appear $\varphi^s(x)_n$, $\varphi^{\sigma}(x)_n$, $\varphi^e(x)_n$, $\varphi^{\varepsilon}(x)_n$.

I. 1.4 Specific Stress Power

On the basis of the preceding relations the specific stress power in a unit volume of the n-th material constituent can be expressed as follows /A- model/:

$$< \sigma_{ij}(\mathbf{X},\mathbf{x},t)_n \, \dot{\varepsilon}_{ij}(\mathbf{X},\mathbf{x},t)_n >_n = \qquad (15)$$

$$= < s_{ij}(\mathbf{X},\mathbf{x},t)_n \, \dot{e}_{ij}(\mathbf{X},\mathbf{x},t)_n + 3\sigma(\mathbf{X},\mathbf{x},t)_n \, \dot{\varepsilon}(\mathbf{X},\mathbf{x},t)_n >_n =$$

$$= < [s^{*}_{ijn}(\mathbf{X},t) + f^{s}(\mathbf{x})_n \, s'_{ijn}(\mathbf{X},t)][\dot{\bar{e}}_{ij}(\mathbf{X},t) + f^{e}(\mathbf{x})_n \, \dot{e}'_{ijn}(\mathbf{X},t)] +$$

$$+ 3[\sigma^{*}_{n}(\mathbf{X},t) + f^{\sigma}(\mathbf{x})_n \, \sigma'_n(\mathbf{X},t)][\dot{\bar{\varepsilon}}(\mathbf{X},t) + f^{\varepsilon}(\mathbf{x})_n \, \dot{\varepsilon}'(\mathbf{X},t)] >_n =$$

$$= s_{ijn}(\mathbf{X},t) \, \dot{e}_{ijn}(\mathbf{X},t) + 3\,\sigma_n(\mathbf{X},t) \, \dot{\varepsilon}_n(\mathbf{X},t) +$$

$$+ \frac{1}{\eta_n} s'_{ijn}(\mathbf{X},t) \, \dot{e}'_{ijn}(\mathbf{X},t) + \frac{3}{\eta^{o}_n} \sigma'_n(\mathbf{X},t) \, \dot{\varepsilon}'_n(\mathbf{X},t)$$

where we have used the relations (11) and (12) and new definitions:

$$\frac{1}{\eta_n} = < [f^{s}(\mathbf{x})_n - 1][f^{e}(\mathbf{x})_n - 1] >_n = \qquad (16)$$

$$= < f^{s}(\mathbf{x})_n \; f^{e}(\mathbf{x})_n >_n - 1$$

$$\frac{1}{\eta^{o}_n} = < [f^{\sigma}(\mathbf{x})_n - 1][f^{\varepsilon}(\mathbf{x})_n - 1] >_n =$$

$$= < f^{\sigma}(\mathbf{x})_n \; f^{\varepsilon}(\mathbf{x})_n >_n - 1$$

The specific stress power in a unit volume of the composite material is:

$$\overline{\sigma_{ij}\dot{\varepsilon}_{ij}} = \sum_{n=1}^{N} v_n \Big[s_{ijn}(\boldsymbol{X},t)\,\dot{e}_{ijn}(\boldsymbol{X},t) + 3\sigma_n(\boldsymbol{X},t)\,\dot{\varepsilon}_n(\boldsymbol{X},t) + \tag{17}$$

$$+ \frac{1}{\eta_n} s'_{ijn}(\boldsymbol{X},t)\,\dot{e}'_{ijn}(\boldsymbol{X},t) + \frac{3}{\eta_n^0}\sigma'_n(\boldsymbol{X},t)\,\dot{\varepsilon}'_n(\boldsymbol{X},t)\Big]$$

Thus, the specific stress power is expressed in terms of functions of macroscopic coordinates only. In any material constituent the expression for the specific stress power is composed of the part that is related to the average values of stress and strain plus another part that describes the influence of fluctuations.

In the case of elasticity the last expression can be integrated and we arrive at the formula for the elastic energy:

$$W^{el} = \frac{1}{2}\sum_{n=1}^{N} v_n \Big[\mu_n \Big(s_{ijn}\, s_{ijn} + \frac{1}{\eta_n} s'_{ijn}\, s'_{ijn} \Big) + \tag{18}$$

$$+ 3\varrho_n \Big(\sigma_n^2 + \frac{1}{\eta_n^0} \sigma_n'^2 \Big) \Big] =$$

$$= \frac{1}{2}\sum_{n=1}^{N} v_n \Big[\frac{1}{\mu_n} \Big(e^{el}_{ijn}\, e^{el}_{ijn} + \frac{1}{\eta_n} e'^{el}_{ijn}\, e'^{el}_{ijn} \Big) +$$

$$+ 3\frac{1}{\varrho_n} \Big(\{\varepsilon_n^{el}\}^2 + \frac{1}{\eta_n^0} \{\varepsilon_n'^{el}\}^2 \Big) \Big]$$

where μ_n, ϱ_n are elastic constants:

$$\mu_n = (1+\nu_n)/E_n\,, \qquad \varrho_n = (1-2\nu_n)/E_n$$

ν_n, E_n being respectively Poisson's ratio and Young's modulus of the n-th material constituent.

In the case of the dual model B the resulting formulae of this paragraph can be rewritten with the only change that instead of η_n, η_n^0 we have to write χ_n, χ_n^0 with the definition:

$$\frac{1}{\chi_n} = \big\langle [\varphi^s(\boldsymbol{x})_n - 1][\varphi^e(\boldsymbol{x})_n - 1] \big\rangle_n = \tag{19}$$

$$= \big\langle \varphi^s(\boldsymbol{x})_n \, \varphi^e(\boldsymbol{x})_n \big\rangle_n - 1$$

$$\frac{1}{\chi_n^o} = \left\langle [\varphi^\sigma(\mathbf{x})_n - 1][\varphi^\varepsilon(\mathbf{x})_n - 1] \right\rangle_n =$$

$$= \left\langle \varphi^\sigma(\mathbf{x})_n \; \varphi^\varepsilon(\mathbf{x})_n \right\rangle_n - 1$$

For the B- model we will need also another equation that can easily be deduced in a quite analogous way as eq. (17):

$$\overline{\varepsilon_{ij}\,\dot{\sigma}_{ij}} = \sum_{n=1}^{N} v_n \Big[e_{ijn}\,\dot{s}_{ijn} + 3\,\varepsilon_n\,\dot{\sigma}_n + \frac{1}{\chi_n}\,e'_{ijn}\,\dot{s}'_{ijn} + \frac{3}{\chi_n^o}\,\varepsilon'_n\,\dot{\sigma}'_n \Big] \tag{20}$$

I.1.5 The Basic Relations Among Internal Stress Components for the A- Model and Strain Components for the B- Model

Let us now combine equations (5) and (17). In (17) we can consider infinitesimal variations instead of rates of strain and in (5) the deformations can also be infinitesimal. Thus we can write:

$$\bar{\sigma}_{ij}\,\delta\bar{\varepsilon}_{ij} - \overline{\sigma_{ij}\,\delta\,\varepsilon_{ij}} = \bar{s}_{ij}\,\delta\,\bar{e}_{ij} + 3\,\bar{\sigma}\,\delta\,\bar{\varepsilon} - \sum_{n=1}^{N} v_n \Big[s_{ijn}\,\delta\,e_{ijn} + 3\,\sigma_n\,\delta\,\varepsilon_n + \frac{1}{\eta_n}\,s'_{ijn}\,\delta\,e'_{ijn} + \frac{3}{\eta_n^o}\,\sigma'_n\,\delta\,\varepsilon'_n \Big] = 0 \tag{21}$$

On the strength of eqs. (1) and $(12)_{1,2}$ the last equation can be rewritten as:

$$\sum_{n=1}^{N} v_n \Big[\Big(s_{ijn} + \frac{s'_{ijn}}{\eta_n} \Big)\,\delta\,e'_{ijn} + 3\Big(\sigma_n + \frac{\sigma'_n}{\eta_n^o} \Big)\,\delta\,\varepsilon'_n \Big] = 0 \tag{22}$$

The variations $\delta e'_{ijn}$ and $\delta \varepsilon'_n$ are not independent, as eqs. (2) and $(12)_1$ give:

$$\sum_{n=1}^{N} v_n \, \delta e'_{ijn} = 0 \tag{23}$$

$$\sum_{n=1}^{N} v_n \, \delta \varepsilon'_n = 0$$

From these equations one arbitrarily chosen variation $\delta e'_{ijm}$ and one $\delta \varepsilon'_m$ can be expressed and the expression used in (22). This leads to:

$$\sum_{n=1,2,---,m-1,m+1,---,N} v_n \left[\left(s_{ijn} - s_{ijm} + \frac{s'_{ijn}}{\eta_n} - \frac{s'_{ijm}}{\eta_m} \right) \delta e'_{ijn} + \right.$$

$$\left. + 3 \left(\sigma_n - \sigma_m + \frac{\sigma'_n}{\eta^0_n} - \frac{\sigma'_m}{\eta^0_m} \right) \delta \varepsilon'_n \right] = 0 \tag{24}$$

Furthermore, the variations $\delta e'_{ijn}$ $(n \neq m)$ are bound by the relation resulting from their deviatoric character:

$$\delta_{ij} \, \delta e'_{ijn} = 0 \tag{25}$$

/ δ_{ij} – Kronecker's delta/.
However, the expression in the brackets preceding $\delta e'_{ijn}$ is also deviatoric and therefore it holds:

$$\delta_{ij} \left(s_{ijn} - s_{ijm} + \frac{s'_{ijn}}{\eta_n} - \frac{s'_{ijm}}{\eta_m} \right) = 0 \tag{26}$$

Let us multiply the last expression by $v_n \lambda_n$, where λ_n are random multipliers, different for different n . Then we add the resulting expression to (24) and we arrive at:

$$\sum_{n=1,2,---,m-1,m+2,---,N} v_n \left[\left(s_{ijn} - s_{ijm} + \frac{s'_{ijn}}{\eta_n} - \frac{s'_{ijm}}{\eta_m} \right) \left(\delta e'_{ijn} + \lambda_n \delta_{ij} \right) + \right.$$

$$\left. + 3 \left(\sigma_n - \sigma_m + \frac{\sigma'_n}{\eta^0_n} - \frac{\sigma'_m}{\eta^0_m} \right) \delta \varepsilon'_n \right] = 0 \tag{27}$$

In the last equation the variations $\delta \varepsilon'_n$ $(n \neq m)$ are independent - arbitrary scalars. The expression

$$\delta e'_{ijn} + \lambda_n \delta_{ij}$$

is an arbitrary tensor, as $\delta e'_{ijn}$ is an arbitrary deviator and λ_n is an arbitrary scalar.

Therefore the following equations hold true:

$$s_{ijn} - s_{ijm} + \frac{s'_{ijn}}{\eta_n} - \frac{s'_{ijm}}{\eta_m} = 0 \tag{28$_1$}$$

$$\sigma_n - \sigma_m + \frac{\sigma'_n}{\eta^o_n} - \frac{\sigma'_m}{\eta^o_m} = 0$$

It is clear enough that the number of the independent equations $(28)_1$ is $N-1$ for the deviatoric parts and $N-1$ for the isotropic parts.

Equations $(28)_1$ are the basic relations among internal stress components for model A.

By a quite analogous procedure the respective basic relations can be deduced for the dual B- model. In this case it is necessary to work with strains and with variations of stresses. Instead of eqs. (17), (1), $(12)_{1,2}$ we use eqs. (20), (2), $(13)_{1,2}$ and arrive at:

$$e_{ijn} - e_{ijm} + \frac{e'_{ijn}}{\chi_n} - \frac{e'_{ijm}}{\chi_m} = 0 \tag{28$_2$}$$

$$\varepsilon_n - \varepsilon_m + \frac{\varepsilon'_n}{\chi^o_n} - \frac{\varepsilon'_m}{\chi^o_m} = 0$$

These are the basic relations among internal strain components for model B. Compared with $(28)_1$ we see that stresses and η are replaced by strains and χ.

I. 1. 6 Constitutive Equations of Material Constituents

It is assumed that for any n- material constituent the respective constitutive equation can be expressed as a special case of the following general formula

$$D_n^{(1)} e_{ij}(X, x, t)_n + D_n^{(2)} \dot{e}_{ij}(X, x, t)_n = \tag{29}$$
$$= D_n^{(3)} s_{ij}(X, x, t)_n + D_n^{(4)} \dot{s}_{ij}(X, x, t)_n$$

$$I_n^{(1)} \varepsilon(\boldsymbol{X}, \boldsymbol{x}, t)_n + I_n^{(2)} \dot{\varepsilon}(\boldsymbol{X}, \boldsymbol{x}, t) =$$

$$= I_n^{(3)} \sigma(\boldsymbol{X}, \boldsymbol{x}, t)_n + I_n^{(4)} \dot{\sigma}(\boldsymbol{X}, \boldsymbol{x}, t)_n +$$

$$+ I_n^{(5)} T(\boldsymbol{X}, t) + I_n^{(6)} \dot{T}(\boldsymbol{X}, t) +$$

$$+ I_n^{(7)} \tau_n(\boldsymbol{X}, t) + I_n^{(8)} \dot{\tau}_n(\boldsymbol{X}, t)$$

where T is temperature and τ_n some other parameter that is specific for the n-th material constituent.

Eq. (29) includes as special cases e.g. thermoelasticity ($D_n^{(1)}$, $D_n^{(3)}$, $I_n^{(1)}$, $I_n^{(3)}$, $I_n^{(5)}$ are constants, the other material parameters have zero values), viscosity, viscoelasticity, plasticity etc. Accordingly the parameters D and I have different meanings. They can be constants or functions of $\boldsymbol{X}, t$ or T. It is a restriction of our model that they cannot be functions of $\boldsymbol{x}$. /Thus e.g. - for fixed $\boldsymbol{X}$ and t - in a phenomenological theory the scalar measure of plastic deformation $d\lambda$ is unique. In our model it can be different in different material constituents, but the fluctuations of plastic deformation are described by the fluctuations of s_{ij} only./

If we perform the averaging procedure $\langle --- \rangle_n$ in eqs. (29), we get:

$$D_n^{(1)} e_{ijn} + D_n^{(2)} \dot{e}_{ijn} = D_n^{(3)} s_{ijn} + D_n^{(4)} \dot{s}_{ijn} \tag{30}$$

$$I_n^{(1)} \varepsilon_n + I_n^{(2)} \dot{\varepsilon}_n = I_n^{(3)} \sigma_n + I_n^{(4)} \dot{\sigma}_n +$$

$$+ I_n^{(5)} T + I_n^{(6)} \dot{T} + I_n^{(7)} \tau_n + I_n^{(8)} \dot{\tau}_n$$

where all the variables are functions of $\boldsymbol{X}$ and t only.

From the definition of s_{ijn}^* and σ_n^* it follows:

$$D_n^{(1)} \bar{e}_{ij} + D_n^{(2)} \dot{\bar{e}}_{ij} = D_n^{(3)} s_{ijn}^* + D_n^{(4)} \dot{s}_{ijn}^* \tag{31}$$

$$I_n^{(1)}\bar{\varepsilon} + I_n^{(2)}\dot{\bar{\varepsilon}} = I_n^{(3)}\sigma_n^* + I_n^{(4)}\dot{\sigma}_n^* +$$

$$+ I_n^{(5)}T + I_n^{(6)}\dot{T} + I_n^{(7)}\tau_n + I_n^{(8)}\dot{\tau}_n$$

Subtraction of (30) and (31) gives /according to (12)/ :

$$D_n^{(1)} e'_{ijn} + D_n^{(2)} \dot{e}'_{ijn} = D_n^{(3)} s'_{ijn} + D_n^{(4)} \dot{s}'_{ijn} \tag{32}$$

$$I_n^{(1)} \varepsilon'_n + I_n^{(2)} \dot{\varepsilon}'_n = I_n^{(3)} \sigma'_n + I_n^{(4)} \dot{\sigma}'_n$$

With the use of (14) eq. (29) can be rewritten as:

$$D_n^{(1)} \bar{e}_{ij} + D_n^{(2)} \dot{\bar{e}}_{ij} + f^e(X)_n \left[D_n^{(1)} e'_{ijn} + D_n^{(2)} \dot{e}'_{ijn} \right] = \tag{33}$$

$$= D_n^{(3)} s^*_{ijn} + D_n^{(4)} \dot{s}^*_{ijn} + f^s(X)_n \left[D_n^{(3)} s'_{ijn} + D_n^{(4)} \dot{s}'_{ijn} \right]$$

$$I_n^{(1)} \bar{\varepsilon} + I_n^{(2)} \dot{\bar{\varepsilon}} + f^{\varepsilon}(X)_n \left[I_n^{(1)} \varepsilon'_n + I_n^{(2)} \dot{\varepsilon}'_n \right] =$$

$$= I_n^{(3)} \sigma_n^* + I_n^{(4)} \dot{\sigma}_n^* + f^{\sigma}(X)_n \left[I_n^{(3)} \sigma'_n + I_n^{(4)} \dot{\sigma}'_n \right] +$$

$$+ I_n^{(5)} T + I_n^{(6)} \dot{T} + I_n^{(7)} \tau_n + I_n^{(8)} \dot{\tau}_n$$

Comparison of (31), (32) , (33) leads to:

$$f^e(X)_n = f^s(X)_n = f(X)_n \tag{34}$$

$$f^{\varepsilon}(X)_n = f^{\sigma}(X)_n = f^0(X)_n$$

where we have newly defined $f(X)_n$ as the distribution function for deviatoric parts and $f^0(X)_n$ as the distribution function for isotropic parts.

From (33) and (16) we conclude:

$$\frac{1}{\eta_n} = \left\langle \left[f(X)_n \right]^2 \right\rangle_n - 1 = \left\langle \left[f(X)_n - 1 \right]^2 \right\rangle_n \geq 0 \tag{35}$$

$$\frac{1}{\eta_n^0} = \langle [f^0(x_n)]^2 \rangle_n - 1 = \langle [f^0(x_n) - 1]^2 \rangle_n \geq 0$$

The non- negative character of the structural parameters η_n, η_n^0 is their very important property and their vanishing and infinite values have special physical meaning.

In the case of the B- model the reasoning is quite similar, eqs. (30) and (32) are formally the same and structural parameters χ_n, χ_n^0 are again non- negative. The respective formulae follow from (35) writing χ, φ instead of η, f.

I.1.7 The Complete Set of Basic Equations

In the preceding paragraphs we have deduced a number of equations that follow from the fundamental theorem of the concept and that form the basis of the analysis. We are going to revise the complete set of these basic independent equations and to show that their number equals the number of the unknowns. For some given instant in the deformation process we assume the stress and strain tensors to be known, the rate of macroscopic stress $\dot{\sigma}_{ij}$ to be given and the rates of the remaining tensor variables are to be determined. The rates of the remaining tensor variables are: $\dot{\sigma}_{ijn}$, $\dot{\sigma}'_{ijn}$, $\dot{\varepsilon}_{ijn}$, $\dot{\varepsilon}'_{ijn}$ and $\dot{\mathcal{E}}_{ij}$. Hence, their number is $4N+1$.

For model A the set of basic equations is represented by:

eq. (1) - 1
eq. (2) - 1
eqs. $(12)_{1,2}$ - N
eqs. $(28)_1$ - $N-1$
eqs. (30) - N
eqs. (32) - N

Thus, the number of basic equations is also $4N+1$. Eqs. (1), (2), $(12)_{1,2}$, $(28)_1$ are written for tensors and not for their rates, but this is only richer information, the analogous equations for rates can be obtained by a trivial derivation.

For model B the set of basic equation is quite similar, instead of $(12)_{1,2}$, $(28)_1$ it is necessary to introduce eqs. $(13)_{1,2}$, $(28)_2$.

From the structure of the basic equations we see that there appear neither the distribution functions f, φ, nor the variables σ^*_{ijn}, ε^*_{ijn}. They served only for the deduction.

The user of the model does not need to know them. Hence, in the final form of the model the restriction imposed is the time-independence only of the structural parameters, which are integral forms of the distributions. Such a restriction is weaker than the time-independence of the distribution functions themselves.

With concrete forms of the constitutive equations of the material constituents in (30) and (32) it is possible to derive the macroscopic constitutive equation. It will be characterized by the parameters of the constitutive equations of the material constituents, their volume fractions, the structural parameters η or $\mathcal{X}$ and generally by internal variables represented by $\sigma_{ijn}, \sigma'_{ijn}$ or $\varepsilon_{ijn}, \varepsilon'_{ijn}$.

I.1.8 The Complete System of Combinations of Compact and Loose Infrastructures of the Material Constituents.

In what follows we will show that with different values of η and $\mathcal{X}$ it is possible to model compact and loose infrastructures of the material constituents. The analysis will be limited to such cases only, where the following quadratic forms /appearing in (18)/ are finite:

$$Q_{sn} = s_{ijn} s_{ijn} + \frac{1}{\eta_n} s'_{ijn} s'_{ijn}$$

$$Q_{\sigma n} = \sigma_n^2 + \frac{1}{\eta_n^0} \sigma_n'^2$$

$$Q_{en} = e_{ijn} e_{ijn} + \frac{1}{\eta_n} e'_{ijn} e'_{ijn}$$

$$Q_{\varepsilon n} = \varepsilon_n^2 + \frac{1}{\eta_n^0} \varepsilon_n'^2$$

In the case of the B- model quite analogous quadratic forms with $\mathcal{X}$ instead of η are considered to be finite too.

Successively we will investigate model A with different values of the structural parameters and then model B.

<u>A 1 : Finite parameters η</u>

We are going to prove that this case corresponds to compact infrastructures of the material constituents, i.e. rigidity of one material constituent leads to rigidity of the composite.

Rigidity of the n-th material constituent in deviatoric deformation can be described by the following special forms of eqs. $(30)_1$ and $(32)_1$:

$$e_{ijn} = \mu_n s_{ijn}$$

$$e'_{ijn} = \mu_n s'_{ijn}$$

where μ_n has a vanishing value. As the Q-forms are finite, the values of s_{ijn}, s'_{ijn} must be finite, too. Therefore, $e_{ijn} = e'_{ijn} = 0$, which - compared with $(12)_1$ - means that $\bar{e}_{ij} = 0$. A quite analogous procedure leads to the conclusion that rigidity in the volumetric deformation leads to $\bar{\varepsilon} = 0$. This shows that rigidity of the n-th material constituent results in rigidity of the composite, i.e. the n-th material constituent has a compact infrastructure.

A 2: Infinite values of one pair of the structural parameters η.

$$\eta_n = \eta_n^0 = \infty$$

In this case eq. $(28)_1$ degenerates to:

$$s_{ijn} - s_{ijm} - \frac{s'_{ijm}}{\eta_m} = 0 \qquad (36)$$

Let the resistance of the m-th material be vanishing. Then its deviatoric properties can be described as follows:

$$s_{ijm} = \frac{1}{\mu_m} e_{ijm}$$

$$s'_{ijm} = \frac{1}{\mu_m} e'_{ijm}$$

where μ_m is infinite. From the value of the Q_{em} -form being limited the values of e_{ijm}, e'_{ijm} follow to be also limited. Therefore, $s_{ijm} = s'_{ijm} = 0$ and this results - with regard to (36) - in $s_{ijn} = 0$.

Thus, the average value s_{ijn} in the n-th material constituent is zero. However, according to $(35)_1$ it holds for infinite η_n :

$$\langle [f(x)_n]^2 \rangle_n = 1$$

By definition

$$\langle f(x)_n \rangle_n = 1$$

and therefore

$$f(x)_n \equiv 1 \quad \text{/cf. Appendix 1/.}$$

This means that the vanishing modulus in the m-th material constituent results in vanishing stress not only in the m-th constituent, but also in the n-th constituent. Therefore, the n-th material constituent is not compact, it is loose. The deduction for the isotropic component would be quite similar. The m-th material constituent is compact, which can be proved in the same way as in A 1.

In a two-phase material, which is the main area of application of our concept, the infinite values of η_n and η_n^o represent the model of inclusions of the n-material in the compact matrix of the m-material.

A 3: Zero values of one pair of the structural parameters η.

$\eta_n = \eta_n^o = 0$

If the structural parameter η_n is zero, then - for the Qsn form to be finite - the value of s'_{ijn} must be also zero. This is valid throughout the deformation process and therefore it also holds $\dot{s}'_{ijn} = 0$. With regard to eq. $(32)_1$ this means that

$$e'_{ijn} = 0 \tag{37}$$

throughout the process /supposing the existence of the virgin state without deformations/.

Further, from (37) and $(14)_1$ we see that

$$e_{ij}(\boldsymbol{X}, \boldsymbol{x}, t)_n = e_{ijn} = \bar{e}_{ij}$$

i.e. deviatoric deformation in the n-material constituent equals the macroscopic deviatoric deformation. The same is valid for the isotropic part of deformation - the deduction is quite analogous.

In the case of a two-phase material it results from (2) that

$$\varepsilon_{ijm} = \varepsilon_{ijn} = \bar{\varepsilon}_{ij}$$

and from $(12)_{1,2}$ and $(14)_{1,2}$:

$$\varepsilon_{ij}(\boldsymbol{X}, \boldsymbol{x}, t)_m = \varepsilon_{ijm} = \bar{\varepsilon}_{ij}$$

Thus in a two-phase material the zero values of one pair of structural parameters η_n, η_n^o are sufficient for the deformation to be homogeneous throughout the composite /Voigt's model/.

A 4: Infinite values of all structural parameters η

In such case all equations $(28)_1$ degenerate to:

$$s_{ijn} = s_{ijm} \; , \; \sigma_n = \sigma_m$$

and by a reasoning analogous to that used in A2 we can see that the stress state in all material constituents is also homogeneous.

With regard to (1) we conclude

$$\sigma_{ij}(X, x, t)_n = \sigma_{ij}(X, x, t)_m = --- = \bar{\sigma}_{ij}$$

i.e. homogeneous stress state throughout the composite /Reuss' solution/.

A 5: Zero values of all structural parameters η :

By a deduction quite similar to that used in A3 we arrive in this case at Voigt's solution, i.e. at homogeneous deformation throughout the composite. If there are N material constituents, it is sufficient if N-1 pairs of structural parameters are zero to arrive at the Voigt solution. The reason was shown in A3 and for a two-phase material models A5 and A3 are equivalent.

B 1: Finite parameters χ

By a procedure that is dual to that used in A1 we are going to prove that in this case vanishing resistance of any of the material constituents leads to vanishing resistance of the composite, which implies that none of the material constituents is compact, i.e. they are all loose.

The complete lack of resistance to deviatoric deformation can be described as:

$$s_{ijn} = e_{ijn} / \mu_n$$
$$s'_{ijn} = e'_{ijn} / \mu_n$$

where μ_n is infinite. Finite Q_{en} implies finite e_{ijn} and e'_{ijn}. Therefore, $s_{ijn} = s'_{ijn} = 0$, which - compared with $(13)_1$ - means that $\bar{s}_{ij} = 0$. A quite analogous procedure leads to $\bar{\sigma} = 0$. This proves that the infrastructures described by model B with finite structural parameters are loose.

B 2: Infinite values of one pair of the structural parameters χ $\chi_n = \chi^o_n = \infty$

In this case eq. $(28)_2$ degenerates to:

$$e_{ijn} - e_{ijm} - \frac{e'_{ijm}}{\chi_m} = 0$$

Let the m-th material be quite rigid, i.e. with infinite Young's modulus. Then /cf. A1/, $e_{ijm} = e'_{ijm} = 0$ and therefore

$$e_{ijn} = 0$$

Using the same reasoning as that used in A2 for $s_{ijn} = 0$, we can easily prove that

$$\varepsilon_{ij}(X, x, t)_n = e_{ijn} = \varepsilon_{ij}(X, x, t)_m = 0$$

For a two-phase material this also means

$$\bar{\varepsilon}_{ij} = 0$$

and therefore, the m-th material constituent has a compact infrastructure in this case.
On the other hand the n-th material constituent has a loose infrastructure, which can be proved in the same way as in B1; here we let the m-th constituent be without resistance and show that this leads to

$$\sigma_{ij}(X, x, t)_n = \sigma_{ij}(X, x, t)_m = \bar{\sigma}_{ij} = 0$$

Thus, for a two-phase material the two models A2 and B2 are equivalent, which will be demonstrated later on concrete examples.

B 3: Zero values of one pair of the structural parameters χ. $\chi_n = \chi_n^o = 0$

We can proceed in a way that is dual to A3, which leads to

$$e'_{ijn} = \dot{e}_{ijn} = 0 \,, \quad s'_{ijn} = 0$$
$$s_{ij}(X, x, t)_n = s_{ijn} = \bar{s}_{ij}$$

and similarly for isotropic parts.

In the case of a two-phase material this results in:

$$\sigma_{ij}(X, x, t)_m = \sigma_{ij}(X, x, t)_n = \bar{\sigma}_{ij}$$

i.e. in the Reuss solution. Thus for a two-phase material model B3 is equivalent to model A4.

B 4: Infinite values of all structural parameters χ.

By a procedure that is dual to that used in A4 we can easily show that in this case:

$$\varepsilon_{ij}(X, x, t)_n = \varepsilon_{ij}(X, x, t)_m = ---- = \bar{\varepsilon}_{ij}$$

i.e. Voigt's solution. Model B4 is therefore equivalent to A5 and in the case of a two-phase material also to A3.

B 5: Zero values of all structural parameters χ.

Similarly as in B3 and A3 we can prove that in this case

it holds:

$$\sigma_{ij}(X, x, t)_n = \sigma_{ij}(X, x, t)_m = ----- = \bar{\sigma}_{ij}$$

i.e. the Reuss solution. Model B5 is therefore equivalent to A4.

Concluding this paragraph let us mention that not all of the possible abstract combinations have physical meaning. The situation is simple and clear in the case of a two-phase material. Models A1, B1 and two equivalent models A2 and B2 have physical meaning of compact infrastructures, loose infrastructures and of inclusions in a matrix. The remaining models correspond to the Voigt and Reuss solutions, which are mathematical fictions only, because there does not exist any threedimensional structure that would lead to homogeneous stress or strain. In spite of it they are used sometimes as simple approximations.

In the case of three material constituents the situation is more complicated. From the preceding deduction it is easy to see that for all η finite we will get a model for three compact infrastructures, for all χ finite a model for three loose infrastructures. For two pairs of η finite and one infinite we get two infrastructures compact and one loose /with contact surfaces with both the compact constituents/, for one pair of η finite and two pairs infinite one infrastructure compact and two loose with equal homogeneous stress in the two loose constituents, for one pair of χ finite and two pairs infinite one infrastructure compact and two loose with equal homogeneous strain in the two loose constituents. Other combinations can be derived similarly. From the above examples we see that in the case of three constituents there arise restrictions and the scheme is not so simple as in a two-phase material. The solution to the identification problem is even more complicated.

I.1.9 An Important Example of a Two-Phase Model

A number of interesting results can be arrived at as special cases of the following pretty general two-phase model. One of the two material constituents is supposed to remain elastic throughout the deformation process /subscript e/ whereas the other material constituent undergoes some kind of inelastic deformation /subscript n/.

I.1.9.1 Compact infrastructures

For the A-model the complete set of basic equations /cf. I.1.7/ reads in our case:

$$v_e \sigma_{ije} + v_n \sigma_{ijn} = \bar{\sigma}_{ij} \tag{38}$$

$$v_e \varepsilon_{ije} + v_n \varepsilon_{ijn} = \bar{\varepsilon}_{ij} \tag{39}$$

$$\varepsilon'_{ije} = \varepsilon_{ije} - \bar{\varepsilon}_{ij} \tag{40}$$

$$\varepsilon'_{ijn} = \varepsilon_{ijn} - \bar{\varepsilon}_{ij}$$

$$s_{ije} - s_{ijn} + \frac{s'_{ije}}{\eta_e} - \frac{s'_{ijn}}{\eta_n} = 0 \tag{41}$$

$$\sigma_e - \sigma_n + \frac{\sigma'_e}{\eta^o_e} - \frac{\sigma'_n}{\eta^o_n} = 0$$

$$e_{ije} = \mu_e s_{ije} \,, \qquad \varepsilon_e = g_e \sigma_e + \alpha_e T \tag{42}$$

$$e'_{ije} = \mu_e s'_{ije} \,, \qquad \varepsilon'_e = g_e \sigma'_e \tag{43}$$

$$\dot{e}_{ijn} = \mu_n \dot{s}_{ijn} + \dot{h} s_{ijn} \,, \quad \dot{\varepsilon}_n = g_n \dot{\sigma}_n + \dot{g}_n \sigma_n + \alpha_n \dot{T} + \dot{\tau} \tag{44}$$

$$\dot{e}'_{ijn} = \mu_n \dot{s}'_{ijn} + \dot{h} s'_{ijn} \,, \quad \dot{\varepsilon}'_n = g_n \dot{\sigma}'_n + \dot{g}_n \sigma'_n \tag{45}$$

Thus, we have got 9 tensorial equations for 9 unknown tensorial rates: $\dot{\bar{\varepsilon}}_{ij}$, $\dot{\varepsilon}_{ije}$, $\dot{\varepsilon}_{ijn}$, $\dot{\varepsilon}'_{ije}$, $\dot{\varepsilon}'_{ijn}$, $\dot{\sigma}_{ije}$, $\dot{\sigma}_{ijn}$, $\dot{\sigma}'_{ije}$, $\dot{\sigma}'_{ijn}$. After a successive elimination procedure /see Appendix 2/ we easily arrive at the respective macroscopic constitutive equation:

$$\dot{\bar{e}}_{ij} = \bar{\mu} \dot{\bar{s}}_{ij} + v_n \left(M s_{ijn} + M' s'_{ijn} \right) \dot{h} \tag{46}$$

$$\dot{\bar{\varepsilon}} = \bar{g} \dot{\bar{\sigma}} + v_n \left(M_o \sigma_n + M'_o \sigma'_n \right) \dot{g}_n + \bar{\alpha} \dot{T} + v_n M_o \dot{\tau}$$

where s_{ijn}, s'_{ijn}, σ_n, σ'_n are tensorial internal variables with evolution equations:

$$\dot{s}_{ijn} = M \dot{\bar{s}}_{ij} - \left(N s_{ijn} - N' s'_{ijn} \right) \dot{h} \tag{47}$$

$$\dot{s}'_{ijn} = \frac{1}{\mu_n}\left[(v_e\mu_n + v_n\mu_e)\dot{s}_{ijn} - \mu_e\dot{\bar{s}}_{ij} + (v_e s_{ijn} - s'_{ijn})\dot{h}\right] =$$

$$= \eta_n\left\{M'\dot{\bar{s}}_{ij} + \left[N's_{ijn} - \frac{\mu_e\eta_e + v_n(v_e\mu_n + v_n\mu_e)}{R}s'_{ijn}\right]\dot{h}\right\}$$

$$\dot{\sigma}_n = M_o\dot{\bar{\sigma}} - (N_o\sigma_n - N'_o\sigma'_n)\dot{\varrho}_n + N_o(\alpha_e - \alpha_n)\dot{T} - N_o\dot{\tau}$$

$$\dot{\sigma}'_n = \frac{1}{\varrho_n}\left[(v_e\varrho_n + v_n\varrho_e)\dot{\sigma}_n - \varrho_e\dot{\bar{\sigma}} + (v_e\sigma_n - \sigma'_n)\dot{\varrho}_n + \right.$$

$$\left. + v_e(\alpha_n - \alpha_e)\dot{T} + v_e\dot{\tau}\right] =$$

$$= \eta^o_n\left\{M'_o\dot{\bar{\sigma}} + \left[N'_o\sigma_n - \frac{\eta^o_e\varrho_e + v_n(v_e\varrho_n + v_n\varrho_e)}{R_o}\sigma'_n\right]\dot{\varrho}_n + \right.$$

$$\left. + (\alpha_n - \alpha_e)N'_o\dot{T} + N'_o\dot{\tau}\right\}$$

and $\bar{\mu}, \bar{\varrho}, \bar{\alpha}, M, M', M_o, M'_o, N, N', N_o, N'_o$ are scalar constants:

$$\bar{\mu} = \mu_e\mu_n\left[v_e\mu_e\eta_e + v_n\mu_n\eta_n + \eta_e\eta_n(v_e\mu_e + v_n\mu_n)\right]/R \tag{48}$$

$$M = \left[\mu_e\mu_n\eta_e\eta_n + \mu_e(v_e\mu_e\eta_e + v_n\mu_n\eta_n)\right]/R \tag{49}$$

$$M' = v_e\mu_e(\mu_n - \mu_e)\eta_e/R$$

$$N = v_e(v_e\mu_e\eta_e + v_n\mu_n\eta_n)/R$$

$$N' = v_e\mu_e\eta_e/R$$

$$R = \mu_e\mu_n\eta_e\eta_n + (v_e\mu_n + v_n\mu_e)(v_e\mu_e\eta_e + v_n\mu_n\eta_n)$$

$$\bar{\varrho} = \varrho_e\varrho_n\left[v_e\varrho_e\eta^o_e + v_n\varrho_n\eta^o_n + \eta^o_e\eta^o_n(v_e\varrho_e + v_n\varrho_n)\right]/R_o \tag{50}$$

$$M_o = \left[\varrho_e\varrho_n\eta^o_e\eta^o_n + \varrho_e(v_e\varrho_e\eta^o_e + v_n\varrho_n\eta^o_n)\right]/R_o \tag{51}$$

$$M_0' = v_e \varrho_e (\varrho_n - \varrho_e) \eta_e^0 / R_0$$

$$N_0 = v_e (v_e \varrho_e \eta_e^0 + v_n \varrho_n \eta_n^0) / R_0$$

$$N_0' = v_e \varrho_e \eta_e^0 / R_0$$

$$R_0 = \varrho_e \varrho_n \eta_e^0 \eta_n^0 + (v_e \varrho_n + v_n \varrho_e)(v_e \varrho_e \eta_e^0 + v_n \varrho_n \eta_n^0)$$

$$\bar{\alpha} = \Big[(v_e \alpha_e + v_n \alpha_n) \varrho_e \varrho_n \eta_e^0 \eta_n^0 + (v_e \alpha_e \varrho_n + v_n \alpha_n \varrho_e) \cdot$$

$$\cdot (v_e \varrho_e \eta_e^0 + v_n \varrho_n \eta_n^0) \Big] / R_0 \tag{52}$$

Inelastic deformation that we describe can be viscous, plastic, of the Bingham type, or caused by quasihomogeneous stable fracturing. Let us suppose for the moment that these four kinds are all present. Then we can write for $\dot{h}$:

$$\dot{h} = \frac{1}{2H} + \dot{\lambda} + \dot{\beta} + \dot{\mu}_n \tag{53}$$

From experience we know that H , $\dot{\lambda}$, $\dot{\beta}$ and $\dot{\mu}_n$ are non-negative / $\dot{\mu}_n$ non-negative means that fracturing cannot cause rise of the shear modulus/. This, however, can be shown from the second law of thermodynamics:

The deviatoric stress-power in a unit volume of the n-material under homogeneous strain is:

$$(\dot{W})_n = s_{ij} \dot{e}_{ij} = \mu_n s_{ij} \dot{s}_{ij} + \dot{h} s_{ij} s_{ij}$$

The rate of the respective elastic energy is:

$$(\dot{E})_n = \frac{d}{dt}\left(\frac{1}{2} \mu_n s_{ij} s_{ij}\right) = \mu_n s_{ij} \dot{s}_{ij} + \frac{1}{2} \dot{\mu}_n s_{ij} s_{ij}$$

The difference of the two expressions represents the rate of energy dissipation that cannot be negative:

$$(W)_n - (\dot{E})_n = s_{ij} s_{ij} \left(\frac{1}{2H} + \dot{\lambda} + \dot{\beta} + \frac{1}{2} \dot{\mu}_n \right) \geq 0 \tag{54}$$

$H, \dot{\lambda}, \dot{\beta}$ and $\dot{\mu}_n$ are independent and therefore:

$$H \geq 0, \quad \dot{\lambda} \geq 0, \quad \dot{\beta} \geq 0, \quad \dot{\mu}_n \geq 0, \quad \dot{h} \geq 0 \tag{55}$$

I.1.9.2 Loose infrastructures

For the B-model equations (38), (39), (42), (43), (44) and (45) will have the same form /although the meaning of σ'_{ije}, σ'_{ijn}, ε'_{ije}, ε'_{ijn} will be different/, but instead of eqs. (40) and (41) we must write:

$$\sigma'_{ije} = \sigma_{ije} - \bar{\sigma}_{ij} \tag{56}$$

$$\sigma'_{ijn} = \sigma_{ijn} - \bar{\sigma}_{ij}$$

$$e_{ije} - e_{ijn} + \frac{e'_{ije}}{\chi_e} - \frac{e'_{ijn}}{\chi_n} = 0 \tag{57}$$

$$\varepsilon_e - \varepsilon_n + \frac{\varepsilon'_e}{\chi^o_e} - \frac{\varepsilon'_n}{\chi^o_n} = 0$$

Again, after a successive elimination procedure /see Appendix 3/ we deduce the respective macroscopic constitutive equation:

$$\dot{\bar{e}}_{ij} = \bar{\mu}\,\dot{\bar{s}}_{ij} + v_n\,(P s_{ijn} + \bar{P}\,\bar{s}_{ij})\,\dot{h} \tag{58}$$

$$\dot{\bar{\varepsilon}} = \bar{g}\,\dot{\bar{\sigma}} + v_n\,(P_o\,\sigma_n + \bar{P}_o\,\bar{\sigma})\,\dot{g}_n + \bar{\alpha}\dot{T} + v_n(P_o + \bar{P}_o)\dot{\tau}$$

where s_{ijn}, σ_n are internal variables with evolution equations:

$$\dot{s}_{ijn} = (P + \bar{P})\,\dot{\bar{s}}_{ij} + (\bar{Q}\,\bar{s}_{ij} - Q\,s_{ijn})\,\dot{h} \tag{59}$$

$$\dot{\sigma}_n = (P_o + \bar{P}_o)\,\dot{\bar{\sigma}} + (\bar{Q}_o\,\bar{\sigma} - Q_o\,\sigma_n)\,\dot{g}_n + (Q_o - \bar{Q}_o)(\alpha_e - \alpha_n)\dot{T} - (Q_o - \bar{Q}_o)\,\dot{\tau}$$

and $\bar{\mu}$, $\bar{g}$, $\bar{\alpha}$, P, $\bar{P}$, Q, $\bar{Q}$, P_o, $\bar{P}_o$, Q_o, $\bar{Q}_o$ are scalar constants:

$$\bar{\mu} = [\mu_e \mu_n \chi_e \chi_n + (v_e \mu_e + v_n \mu_n)(v_e \mu_n \chi_e + v_n \mu_e \chi_n)]/R \tag{60}$$

$$P = \mu_e\,(\chi_e \chi_n + v_e \chi_e + v_n \chi_n)/R \tag{61}$$

$$\bar{P} = v_e\,(\mu_n - \mu_e)\,\chi_e/R$$

$$Q = v_e\,\chi_e\,(1 + \chi_n)/R$$

$$\bar{Q} = v_e\,\chi_e/R$$

$$R = (v_e \mu_n + v_n \mu_e) \chi_e \chi_n + v_e \mu_n \chi_e + v_n \mu_e \chi_n \tag{62}$$

$$\bar{\varrho} = \left[\varrho_e \varrho_n \chi_e^o \chi_n^o + (v_e \varrho_e + v_n \varrho_n)(v_e \varrho_n \chi_e^o + v_n \varrho_e \chi_n^o)\right] / R_o$$

$$\bar{\alpha} = \left[(v_e \alpha_e \varrho_n + v_n \alpha_n \varrho_e) \chi_e^o \chi_n^o + (v_e \alpha_e + v_n \alpha_n)(v_e \varrho_n \chi_e^o + \right.$$
$$\left. + v_n \varrho_e \chi_n^o)\right] / R_o \tag{63}$$

$$P_o = \varrho_e (\chi_e^o \chi_n^o + v_e \chi_e^o + v_n \chi_n^o) / R_o \tag{64}$$

$$\bar{P}_o = v_e (\varrho_n - \varrho_e) \chi_e^o / R_o$$

$$Q_o = v_e \chi_e^o (1 + \chi_n^o) / R_o$$

$$\bar{Q}_o = v_e \chi_e^o / R_o$$

$$R_o = (v_e \varrho_n + v_n \varrho_e) \chi_e^o \chi_n^o + v_e \varrho_n \chi_e^o + v_n \varrho_e \chi_n^o$$

Let us shortly comment on the two dual models A and B:

- To underline the similarity and duality of the two models, we use the same symbols for analogous quantities σ_{ij}', ε_{ij}', $\bar{\mu}$, $\bar{\varrho}$, $\bar{\alpha}$, R, although their definition in the two models is different.
- In the A-model there are two tensorial internal variables, whereas in the B-model only one.
- The choice of σ_{ijn} for the internal variable is not the only possibility. It seems to be advantageous in the case of plastic deformation because of its adequacy for the formulation of the yield condition in the n-material, but in other cases other internal variables can be preferable.
- It is easy to see that the infrastructures described by the A-model with finite structural parameters are really compact /cf. I. 1.8-A1/ and those described by the B-model with finite structural parameters are loose /cf. I. 1.8-B1/. To this purpose let either μ_e, ϱ_e or μ_n, ϱ_n in eqs.

(48),(50) be vanishing /one of the material constituents rigid/. This gives $\bar{\mu}$ and $\bar{\varrho}$ vanishing too. However, it is not the case in eqs. (60), (62). On the other hand, if μ_e, ϱ_e or μ_n, ϱ_n go to infinity /one of the material constituents without resistance/ in eqs. (60), (62), then $\bar{\mu}, \bar{\varrho}$ go to infinity too. However, this is not the case in eqs. (48), (50). Similarly for ϱ_e, α_e or ϱ_n, α_n vanishing in (52) $\bar{\alpha}$ is vanishing, but not in (63). The influence of rigidity or of the lack of resistance is mentioned only from the point of view of the elastic constants, but it leads to the same conclusion if the inelastic deformation is also taken into account.

I.1.9.3 The elastic inclusions in the inelastic matrix

Let us start with the A-model. According to I.1.8 the case of elastic inclusions is to be described by the A-model with η_e, η_e^0 infinite. For such a case eqs. (46) and (47) can be simplified in their forms. There is no need to have σ'_{ijn} as one of the internal variables, as it can be simply expressed from eq. (41), which leads - due to the infinity of η_e, η_e^0 - to:

$$s'_{ijn} = \eta_n (s_{ije} - s_{ijn}) = -\frac{\eta_n}{v_e}(s_{ijn} - \bar{s}_{ij}) \tag{65}$$

$$\sigma'_n = \eta_n^0 (\sigma_e - \sigma_n) = -\frac{\eta_n^0}{v_e}(\sigma_n - \bar{\sigma})$$

Using the last equations in (46) and (47), we arrive at:

$$\dot{e}_{ij} = \bar{\mu}\,\dot{\bar{s}}_{ij} + v_n (P s_{ijn} + \bar{P}\bar{s}_{ij})\,\dot{h} \tag{66}$$

$$\dot{\bar{\varepsilon}} = \bar{\varrho}\,\dot{\bar{\sigma}} + v_n (P_0 \sigma_n + \bar{P}_0 \bar{\sigma})\,\dot{\varrho}_n + \bar{\alpha}\dot{T} + v_n (P_0 + \bar{P}_0)\,\dot{\tau}$$

$$\dot{s}_{ijn} = (P + \bar{P})\,\dot{\bar{s}}_{ij} + (\bar{Q}\bar{s}_{ij} - Q s_{ijn})\,\dot{h} \tag{67}$$

$$\dot{\sigma}_n = (P_0 + \bar{P}_0)\,\dot{\bar{\sigma}} + (\bar{Q}_0 \bar{\sigma} - Q_0 \sigma_n)\,\dot{\varrho}_n + (Q_0 - \bar{Q}_0)(\alpha_e - \alpha_n)\dot{T} - (Q_0 - \bar{Q}_0)\,\dot{\tau}$$

The coefficients are easily obtained from eqs. (48) to (52) with infinite values of η_e, η_e^0 and with the use of eqs. (65):

$$\bar{\mu} = \mu_n [v_e \mu_e + \eta_n (v_e \mu_e + v_n \mu_n)] / R \tag{68}$$

$$P = \mu_e (v_e + \eta_n) / R \tag{69}$$

$$\bar{P} = \eta_n(\mu_n - \mu_e)/R$$

$$Q = (v_e^2 + \eta_n)/R$$

$$\bar{Q} = \eta_n/R$$

$$R = \mu_n \eta_n + v_e(v_e \mu_n + v_n \mu_e)$$

$$\bar{\varrho} = \varrho_n[v_e \varrho_e + \eta_n^o(v_e \varrho_e + v_n \varrho_n)]/R_o \tag{70}$$

$$\bar{\alpha} = [(v_e \alpha_e + v_n \alpha_n)\varrho_n \eta_n^o + v_e(v_e \alpha_e \varrho_n + v_n \alpha_n \varrho_e)]/R_o \tag{71}$$

$$P_o = \varrho_e(v_e + \eta_n^o)/R_o \tag{72}$$

$$\bar{P}_o = \eta_n^o(\varrho_n - \varrho_e)/R_o$$

$$Q_o = (v_e^2 + \eta_n^o)/R_o$$

$$\bar{Q}_o = \eta_n^o/R_o$$

$$R_o = \varrho_n \eta_n^o + v_e(v_e \varrho_n + v_n \varrho_e)$$

If we go with μ_e, ϱ_e or μ_n, ϱ_n to infinity or to vanishing values, we can show - similarly as it was done at the end of the preceding paragraph - that the e-material forms a loose infrastructure and the n-material a compact infrastructure.

Now let us turn to the B-model. According to I.1.8 the case of elastic inclusions is to be described also by the B-model with χ_e, χ_e^o infinite. For such a case the form of eqs. (58) and (59) remains unchanged and eqs. (60) to (64) change in the following way:

$$\bar{\mu} = \mu_n[\mu_e \chi_n + v_e(v_e \mu_e + v_n \mu_n)]/R \tag{73}$$

$$P = \mu_e(\chi_n + v_e)/R \tag{74}$$

$$\bar{P} = v_e(\mu_n - \mu_e)/R$$

$$Q = v_e(1+\chi_n)/R$$

$$\bar{Q} = v_e/R$$

$$R = (v_e\mu_n + v_n\mu_e)\chi_n + v_e\mu_n$$

$$\bar{\varrho} = \varrho_n\left[\varrho_e\chi_n^0 + v_e(v_e\varrho_e + v_n\varrho_n)\right]/R_0 \tag{75}$$

$$\bar{\alpha} = \left[(v_e\alpha_e\varrho_n + v_n\alpha_n\varrho_e)\chi_n^0 + (v_e\alpha_e + v_n\alpha_n)v_e\varrho_n\right]/R_0 \tag{76}$$

$$P_0 = \varrho_e(\chi_n^0 + v_e)/R_0 \tag{77}$$

$$\bar{P}_0 = v_e(\varrho_n - \varrho_e)/R_0$$

$$Q_0 = v_e(1+\chi_n^0)/R_0$$

$$\bar{Q}_0 = v_e/R_0$$

$$R_0 = (v_e\varrho_n + v_n\varrho_e)\chi_n^0 + v_e\varrho_n$$

Again, it is easy to show that the infrastructure of the e-material - described in this way - is loose and that of the n-material compact. Moreover, the two mathematical models are equivalent. This statement can be proved simply by the following substitution: the symbols η_n, η_n^0 in eqs. (66) to (72) are to be replaced by v_e^2/χ_n, v_e^2/χ_n^0 respectively. This leads to eqs. (58) and (59) with the coefficients defined by (73) to (77).

I. 1.9.4 The inelastic inclusions in the elastic matrix

Again we will start with the A-model. In this case the structural parameters η_n, η_n^0 go to infinity. The forms of eqs. (46) and $(47)_{1,3}$ remain unchanged and eqs. (48) to (52) change as follows:

$$\bar{\mu} = \mu_e\left[v_n\mu_n + \eta_e(v_e\mu_e + v_n\mu_n)\right]/R \tag{78}$$

$$M = \mu_e(v_n + \eta_e)/R \tag{79}$$

$$M' = 0$$

$$N = v_e v_n/R$$

$$N' = 0$$

$$R = \mu_e \eta_e + v_n (v_e \mu_n + v_n \mu_e)$$

$$\bar{\varrho} = \varrho_e \left[v_n \varrho_n + \eta_e^o (v_e \varrho_e + v_n \varrho_n) \right] / R_o \qquad (80)$$

$$M_o = \varrho_e (v_n + \eta_e^o) / R_o \qquad (81)$$

$$M_o' = 0$$

$$N_o = v_e v_n / R_o$$

$$N_o' = 0$$

$$R_o = \varrho_e \eta_e^o + v_n (v_e \varrho_n + v_n \varrho_e)$$

$$\bar{\alpha} = \left[(v_e \alpha_e + v_n \alpha_n) \varrho_e \eta_e^o + (v_e \alpha_e \varrho_n + v_n \alpha_n \varrho_e) v_n \right] / R_o \qquad (82)$$

Because of the vanishing values of M', N', M_o', N_o' the macroscopic constitutive equation (46)-(47) does not comprise σ_{ijn}' as the second internal variable, the only internal variable is here σ_{ijn} .

Again if the values of μ_e, ϱ_e or μ_n, ϱ_n go to infinity or to zero, it is easy to show - similarly as it was done at the end of I. 1.9 - that the e-material forms here a compact infrastructure and the n-material a loose infrastructure.

Now to the B-model:

According to I. 1.8 the case of inelastic inclusions is to be described also by the B-model with χ, χ_n^o infinite. The forms of eqs. (58) and (59) do not change and eqs. (60) to (64) are:

$$\bar{\mu} = \mu_e \left[\mu_n \chi_e + v_n (v_e \mu_e + v_n \mu_n) \right] / R \qquad (83)$$

$$P = \mu_e (\chi_e + v_n) / R \qquad (84)$$

$$\bar{P} = 0$$

$$Q = v_e \chi_e / R$$

$$\bar{Q} = 0$$

$$R = (v_e \mu_n + v_n \mu_e) \chi_e + v_n \mu_e$$

$$\bar{\varrho} = \varrho_e \left[\varrho_n \chi_e^o + v_n (v_e \varrho_e + v_n \varrho_n) \right] / R_o \tag{85}$$

$$\bar{\alpha} = \left[(v_e \alpha_e \varrho_n + v_n \alpha_n \varrho_e) \chi_e^o + (v_e \alpha_e + v_n \alpha_n) v_n \varrho_e \right] / R_o \tag{86}$$

$$P_o = \varrho_e (\chi_e^o + v_n) / R_o \tag{87}$$

$$\bar{P}_o = 0$$

$$Q_o = v_e \chi_e^o / R_o$$

$$\bar{Q}_o = 0$$

$$R_o = (v_e \varrho_n + v_n \varrho_e) \chi_e^o + v_n \varrho_e$$

It can easily be proved that this model also corresponds to a compact infrastructure of the e-material and to a loose infrastructure of the n-material. The vanishing values of $\bar{P}$, $\bar{Q}$, $\bar{P}_o$, $\bar{Q}_o$ indicate that macroscopic stress does not appear in the macroscopic constitutive equations, which means that the general form of eqs. (46) and $(47)_{1,3}$ with definitions (78) to (82) is the same as that of eqs. (58) and (59) with definitions (83) to (87). The exact equivalence of these two sets of equations can be proved if the symbols η_e, η_e^o in eqs. (78) to (82) /defining coefficients to (46) and $(47)_{1,3}$ / are replaced by v_n^2/χ_e, v_n^2/χ_e^o respectively. This leads to eqs. (58) and (59) with definitions (83) to (87).

I.1.9.5 Homogeneous stress model

Supposition of homogeneous stress in the heterogeneous material - often called Reuss' solution - corresponds to a series arrangement of rheological models. According to I.1.8 this trivial variant is to be arrived at either from the A-model with all structural parameters infinite or from the B-model, with one or both pairs of the structural parameters vanishing. It is a straightforward matter ro show that it is really so. The resulting relations are:

$$\dot{\bar{e}}_{ij} = (v_e \mu_e + v_n \mu_n) \dot{\bar{s}}_{ij} + v_n \bar{s}_{ij} \dot{h} \tag{88}$$

$$\dot{\bar{\varepsilon}} = (v_e \varrho_e + v_n \varrho_n) \dot{\bar{\sigma}} + v_n \bar{\sigma} \dot{\varrho}_n + (v_e \alpha_e + v_n \alpha_n) \dot{T} + v_n \dot{\tau}$$

$$\sigma_{ijn} = \sigma_{ije} = \bar{\sigma}_{ij} \tag{89}$$

I.1.9.6 Homogeneous strain model

Supposition of homogeneous strain in the heterogeneous material - often called Voigt's solution - corresponds to a parallel arrangement of rheological models. According to I.1.8 this variant is to be derived either from the A-model with one or both pairs of the structural parameters vanishing, or from the B-model with all structural parameters infinite. From both these ways the unique result is:

$$\dot{\bar{e}}_{ij} = \dot{e}_{ije} = \dot{e}_{ijn} = \frac{\mu_e}{v_e \mu_n + v_n \mu_e} (\mu_n \dot{\bar{s}}_{ij} + v_n s_{ijn} \dot{h}) \tag{90}$$

$$\dot{\bar{\varepsilon}} = \dot{\varepsilon}_e = \dot{\varepsilon}_n = \frac{\varrho_e}{v_e \varrho_n + v_n \varrho_e} \left[\varrho_n \dot{\bar{\sigma}} + v_n \sigma_n \dot{\varrho}_n + v_n \dot{\tau} \right] + \frac{v_e \alpha_e \varrho_n + v_n \alpha_n \varrho_e}{v_e \varrho_n + v_n \varrho_e} \dot{T}$$

$$\dot{s}_{ijn} = \frac{1}{v_e \mu_n + v_n \mu_e} \left[\mu_e \dot{\bar{s}}_{ij} - v_e s_{ijn} \dot{h} \right] \tag{91}$$

$$\dot{\sigma}_n = \frac{1}{v_e \varrho_n + v_n \varrho_e} \left[\varrho_e \dot{\bar{\sigma}} - v_e \sigma_n \dot{\varrho}_n + v_e (\alpha_e - \alpha_n) \dot{T} - v_e \dot{\tau} \right]$$

In eqs. (90) the internal variables s_{ijn}, σ_n can be excluded with the use of eqs. (1) and (42) :

$$v_n s_{ijn} = \bar{s}_{ij} - v_e s_{ije} = \bar{s}_{ij} - v_e \bar{e}_{ij} / \mu_e \tag{92}$$

$$v_n \sigma_n = \bar{\sigma} - v_e \sigma_e = \bar{\sigma} - v_e (\bar{\varepsilon} - \alpha_e T) / \varrho_e$$

I.1.9.7 Two demonstrative schemes: the elastic constituent with the viscous one and the elastic constituent with the rigid-plastic one

To clarify the meaning of the formulae two simplest examples are discussed and demonstrated in figures.

a/ The elastic constituent with the viscous one

Let us demonstrate the mechanical behaviour of a two-phase material with the following special values of the constants in eqs. (42) to (45) :

$$\varrho_e = \alpha_e = \mu_n = \varrho_n = \alpha_n = \dot{\tau} = 0 \tag{93}$$

$$\dot{h} = \frac{1}{2H} \tag{94}$$

which means that the e-material has only deviatoric elastic deformation and the n-material only deviatoric viscous deformation.

It is a straightforward matter to derive the respective forms of the general equations:

- of eqs. (46) and (47) for compact infrastructures:

$$\dot{\bar{e}}_{ij} = \frac{1}{2H} (s_{ijn} - s'_{ijn}) \tag{95}$$

$$\dot{s}_{ijn} = \frac{1}{v_n} \left(\dot{\bar{s}}_{ij} - \frac{v_e s_{ijn} - s'_{ijn}}{2H \mu_e} \right)$$

$$\dot{s}'_{ijn} = \frac{\eta_n}{v_n} \left[-\dot{\bar{s}}_{ij} + \frac{1}{2H \mu_e} \left(s_{ijn} - \frac{\eta_e + v_n^2}{v_e \eta_e} s'_{ijn} \right) \right]$$

or - after elimination of the internal variables:

$$2H \mu_e v_e v_n \eta_e \ddot{\bar{e}}_{ij} + (v_e^2 \eta_e + v_n^2 \eta_n + \eta_e \eta_n) \dot{\bar{e}}_{ij} + \frac{v_e v_n \eta_n}{2H \mu_e} \bar{e}_{ij} =$$

$$= \mu_e v_e \eta_e (1 + \eta_n) \dot{\bar{s}}_{ij} + v_n \eta_n \frac{1 + \eta_e}{2H} \bar{s}_{ij} \tag{95a}$$

- of eqs. (58) and (59) <u>for loose infrastructures:</u>

$$\dot{e}_{ij} = \frac{1}{1+\chi_e}\left\{ v_e \mu_e \dot{\bar{s}}_{ij} + \frac{1}{2H\chi_n}\left[(v_e \chi_e + v_n \chi_n + \chi_e \chi_n) s_{ijn} - \right.\right.$$
$$\left.\left. - v_e \chi_e \bar{s}_{ij} \right]\right\} \tag{96}$$

$$\dot{s}_{ijn} = \frac{v_n + \chi_e}{v_n(1+\chi_e)}\dot{\bar{s}}_{ij} + \frac{v_e \chi_e}{2H\mu_e v_n \chi_n (1+\chi_e)}\left[\bar{s}_{ij} - (1+\chi_n) s_{ijn} \right]$$

or - without internal variables:

$$2H\mu_e v_n \chi_n (1+\chi_e)\ddot{\bar{e}}_{ij} + v_e \chi_e (1+\chi_n)\dot{\bar{e}}_{ij} =$$
$$= 2H\mu_e^2 v_e v_n \chi_n \ddot{\bar{s}}_{ij} + \mu_e (v_e^2 \chi_e + v_n^2 \chi_n + \chi_e \chi_n)\dot{\bar{s}}_{ij} +$$
$$+ \frac{v_e v_n \chi_e}{2H} \bar{s}_{ij} \tag{96a}$$

- of eqs. (66) and (67) <u>for the elastic inclusions in the viscous matrix:</u>

$$\dot{\bar{e}}_{ij} = \frac{1}{2H v_e}\left[(v_e + \eta_n) s_{ijn} - \eta_n \bar{s}_{ij} \right] \tag{97}$$

$$\dot{s}_{ijn} = \frac{1}{v_n}\left\{ \dot{\bar{s}}_{ij} + \frac{1}{2H\mu_e v_e}\left[\eta_n \bar{s}_{ij} - (v_e^2 + \eta_n) s_{ijn} \right]\right\}$$

or - without internal variables:

$$2H\mu_e v_e v_n \ddot{\bar{e}}_{ij} + (v_e^2 + \eta_n)\dot{\bar{e}}_{ij} = \mu_e v_e (1+\eta_n)\dot{\bar{s}}_{ij} +$$
$$+ \frac{v_n \eta_n}{2H} \bar{s}_{ij} \tag{97a}$$

- of eqs. (46) and (47) with definitions (78) and (79) <u>for the viscous inclusions in the elastic matrix:</u>

$$\dot{\bar{e}}_{ij} = \frac{1}{v_n^2 + \eta_e}\left(v_e \mu_e \eta_e \dot{\bar{s}}_{ij} + v_n \frac{v_n + \eta_e}{2H} s_{ijn} \right) \tag{98}$$

$$\dot{s}_{ijn} = \frac{1}{v_n^2 + \eta_e}\left[(v_n + \eta_e)\dot{\bar{s}}_{ij} - \frac{v_e v_n}{2H\mu_e} s_{ijn} \right]$$

or - without internal variables:

$$(v_n^2 + \eta_e)\dot{\bar{e}}_{ij} + \frac{v_e v_n}{2H\mu_e}\bar{e}_{ij} = \mu_e v_e \eta_e \dot{\bar{s}}_{ij} + v_n \frac{1+\eta_e}{2H}\bar{s}_{ij} \tag{98a}$$

- of eq.(88) for the homogeneous stress model /Maxwell's body/:

$$\dot{\bar{e}}_{ij} = v_e \mu_e \dot{\bar{s}}_{ij} + \frac{v_n}{2H} \bar{s}_{ij} \tag{99}$$

$$\dot{s}_{ijn} = \dot{\bar{s}}_{ij}$$

- and of eqs. (90) and (92) for the homogeneous strain model /Kelvin's body/:

$$\dot{\bar{e}}_{ij} = \frac{s_{ijn}}{2H} = \frac{1}{2H v_n}\left(\bar{s}_{ij} - v_e \frac{\bar{e}_{ij}}{\mu_e}\right) \tag{100}$$

$$\dot{s}_{ijn} = \frac{1}{v_n}\left(\dot{\bar{s}}_{ij} - \frac{v_e s_{ijn}}{2H \mu_e}\right)$$

Equations without internal variables (95_a), (96_a), (97_a), (98_a), $(99)_1$, and $(100)_1$ /the last form/ are usually called rheological equations and it is easy to see that they represent six different types. If there appears the zeroth derivative of strain, deformation after unloading starts going back to zero value. In the opposite case the material has the property called "secondary creep", i.e. deformation after unloading will not vanish. In accordance with the physical insight the former case corresponds to compact infrastructures of the elastic constituent, the latter case to loose infrastructures of the elastic constituent.

Furthermore, let us mention that eq. (97_a) corresponds to the type of the rheological equation of the "Lethersich body" /cf. Reiner [63]/ that was found as suitable for the description of the rheological behaviour of elastic sols, i.e. of suspensions of elastic particles in viscous liquid. In our concept this correspondence is straightforward: eq. (97_a) was derived as descriptive of a heterogeneous material formed by a viscous matrix with elastic inclusions.

For a graphic demonstration we suppose constant loading by a uniaxial macroscopic stress $\bar{\sigma}_{11}$ on a limited time-interval. The deviatoric part of the macroscopic stress tensor is:

$$\bar{s}_{ij} = \bar{s}\begin{pmatrix} 1 & 0 & 0 \\ 0 & -0.5 & 0 \\ 0 & 0 & -0.5 \end{pmatrix} \tag{101}$$

The deviatoric matrix on the right hand side of the equation is the same for the macroscopic deviatoric deformation $\bar{e}_{ij}$ as well as for all other deviatoric stresses and strains on the mesoscale. Instead of $\bar{s}$ in front of the matrix we have to write $\bar{e}, s_e, s_n, e_e, e_n$. The courses of stresses and strains on the macroscale and the mesoscale are plotted in the following pictures:

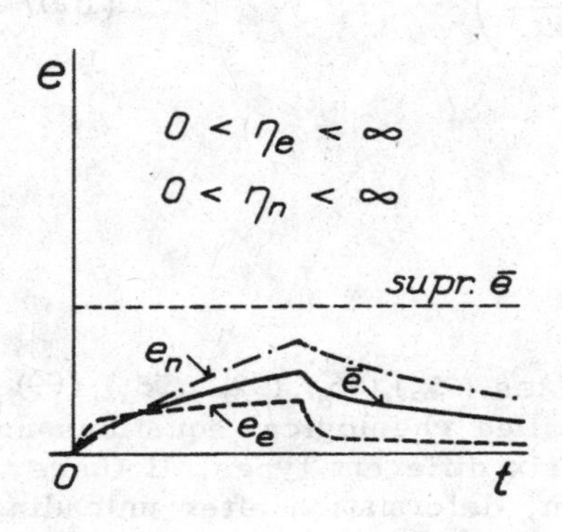

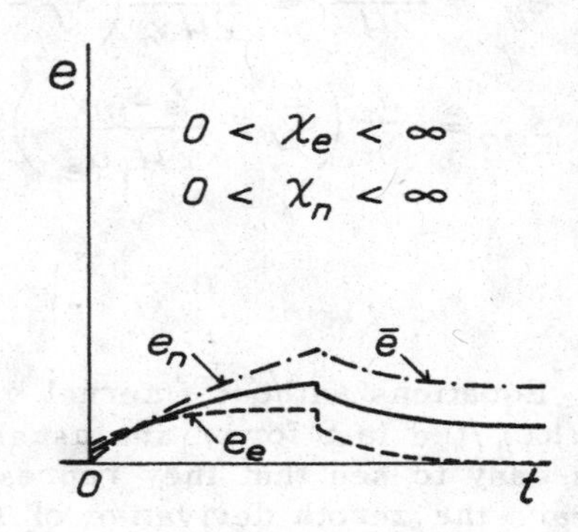

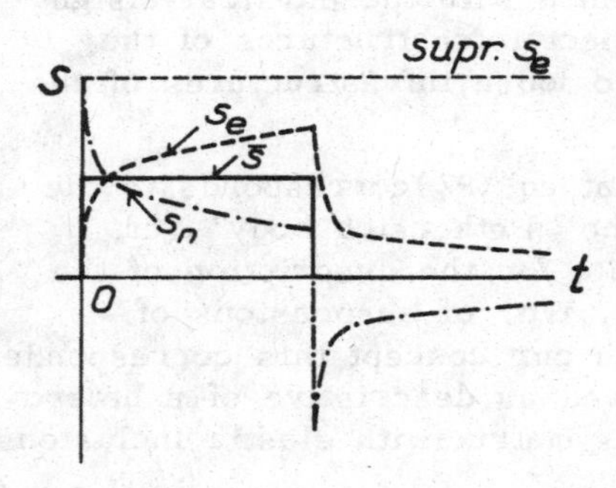

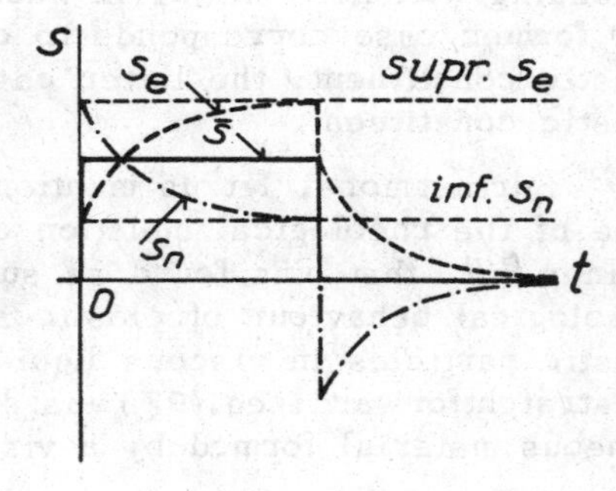

Fig. 1 Compact infrastructures - elastic and viscous constituents

Fig. 2 Loose infrastructures - elastic and viscous constituents

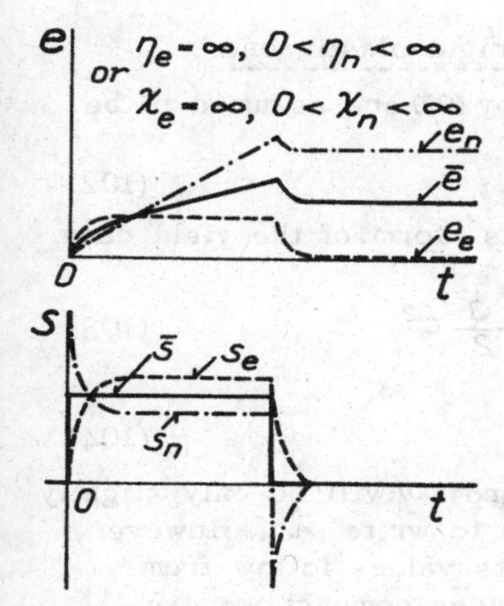

Fig. 3 Elastic inclusions in a viscous matrix

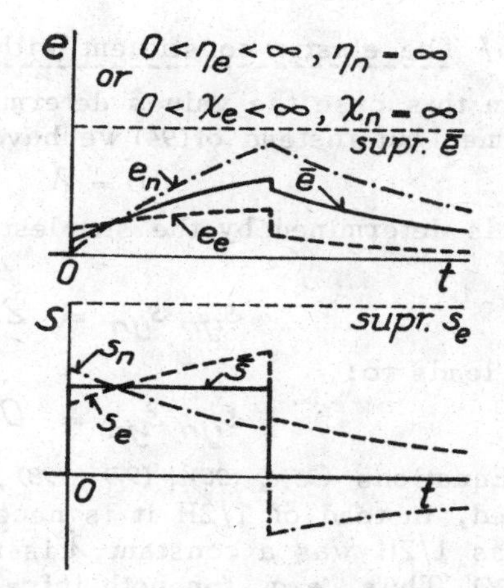

Fig. 4 Viscous inclusions in an elastic matrix

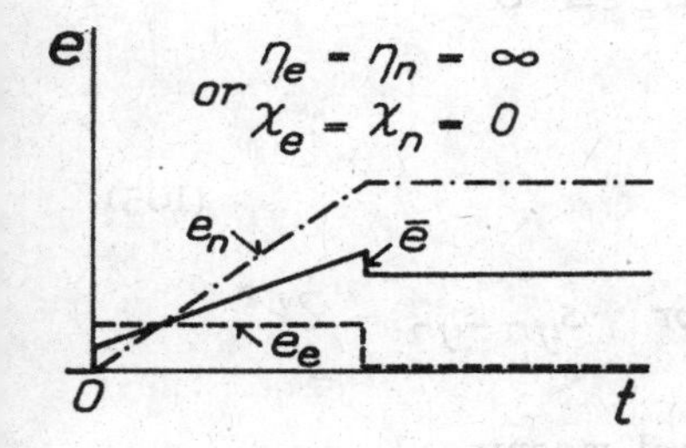

Fig.5 Homogeneous stress model - elastic and viscous constituents /Maxwell's body/

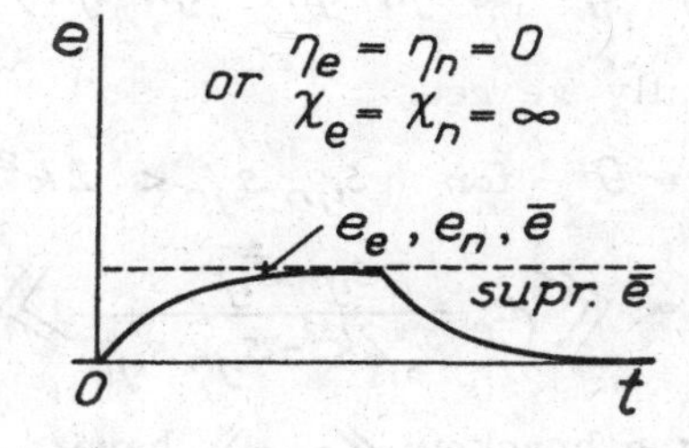

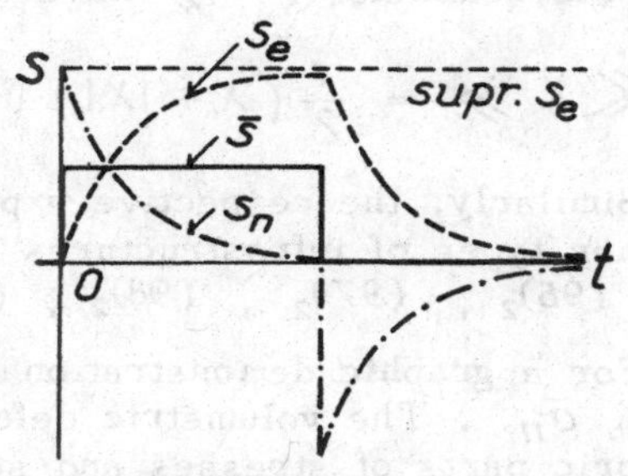

Fig.6 Homogeneous strain model - elastic and viscous constituents /Kelvin's body/

b/ The elastic constituent with the rigid-plastic one

In this case the values determined by (93) are assumed to be the same, but instead of (94) we have:

$$\dot{h} = \dot{\lambda} \tag{102}$$

and $\dot{\lambda}$ is determined by the simplest Mises' form of the yield condition:

$$s_{ijn}\, s_{ijn} = 2k^2 = \frac{3}{2} c^2 \tag{103}$$

which leads to:

$$s_{ijn}\, \dot{s}_{ijn} = 0 \tag{104}$$

Equations (95), (96), (97), (98), (99) and (100) will be only slightly modified, instead of 1/2H it is necessary to write $\dot{\lambda}$. However, whereas 1/2H was a constant, $\dot{\lambda}$ is not. Its values follow from eq. (104). Thus, e.g. for both infrastructures compact we use eqs. $(95)_2$ and (104), which yields:

$$s_{ijn}\, \dot{\bar{s}}_{ij} - (v_e\, s_{ijn}\, s_{ijn} - s_{ijn}\, s'_{ijn})\, \dot{\lambda} / \mu_e = 0$$

and finally we get:

$$\dot{\lambda} = 0 \quad \text{for} \quad s_{ijn}\, s_{ijn} < 2k^2 \tag{105}$$

$$\dot{\lambda} = \left\langle\!\left\langle \mu_e \frac{s_{ijn}\, \dot{\bar{s}}_{ij}}{v_e\, 2k^2 - s_{ijn}\, s'_{ijn}} \right\rangle\!\right\rangle \quad \text{for} \quad s_{ijn}\, s_{ijn} = 2k^2$$

where the brackets $\langle\!\langle --- \rangle\!\rangle$ have a special meaning:

$$\langle\!\langle X \rangle\!\rangle = \frac{1}{2}(X + |X|) \tag{106}$$

Similarly, the respective expressions for $\dot{\lambda}$ in the cases of the other types of infrastructures follow in a straightforward way from $(96)_2$, $(97)_2$, $(98)_2$, $(99)_2$ and $(100)_2$.

For a graphic demonstration we suppose increasing uniaxial tension $\bar{\sigma}_{11}$. The volumetric deformation is again zero and the deviatoric parts of stresses and strains on the macroscale as well as on the mesoscale are given by a variable scalar and the constant matrix that is shown in eq. (101). The respective courses of the scalars $\bar{s}, \bar{e}, s_e, s_n, e_e, e_n$ are given in figures 7 to 12.

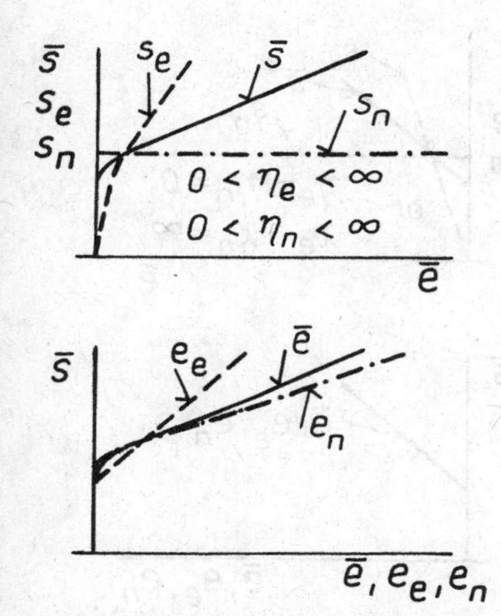

Fig. 7 Compact infrastructures - elastic and rigid-plastic constituents

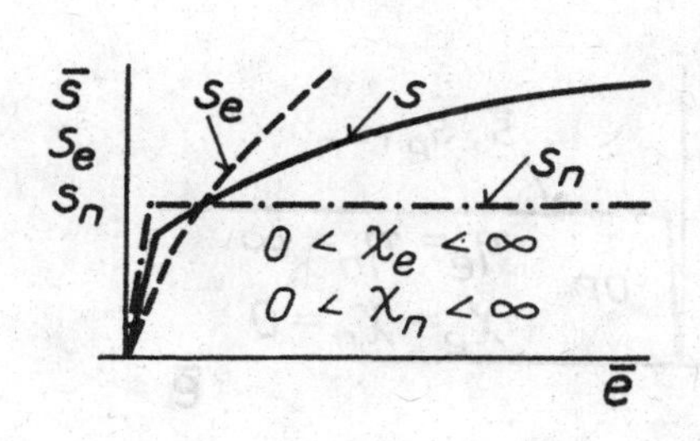

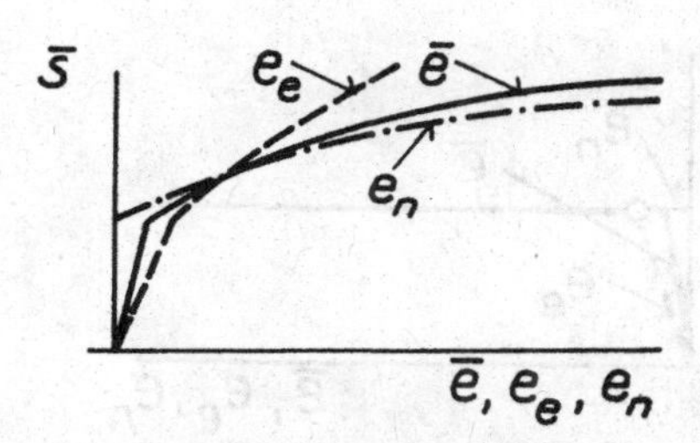

Fig. 8 Loose infrastructures - elastic and rigid-plastic constituents

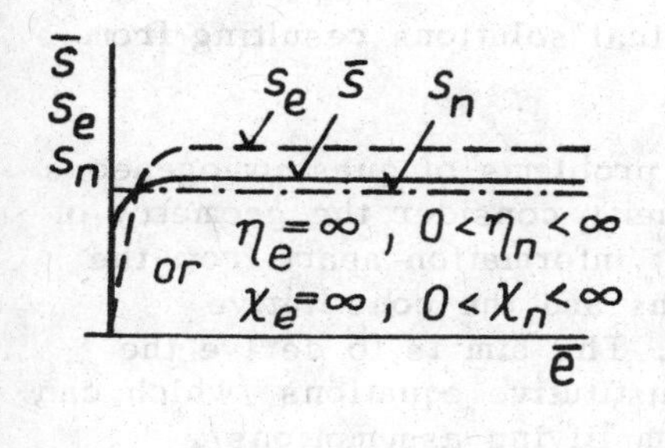

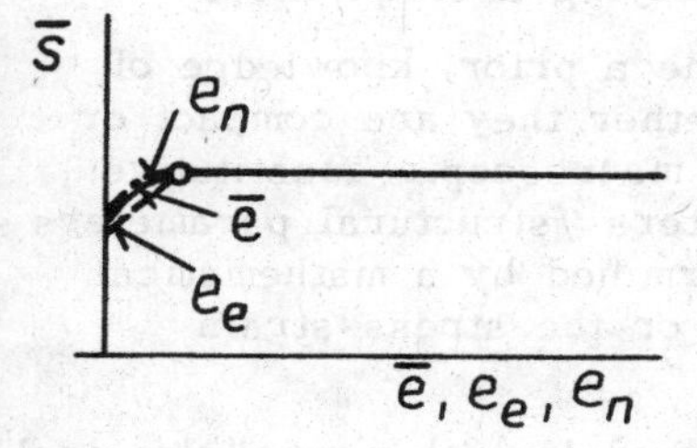

Fig. 9 Elastic inclusions in a rigid-plastic matrix

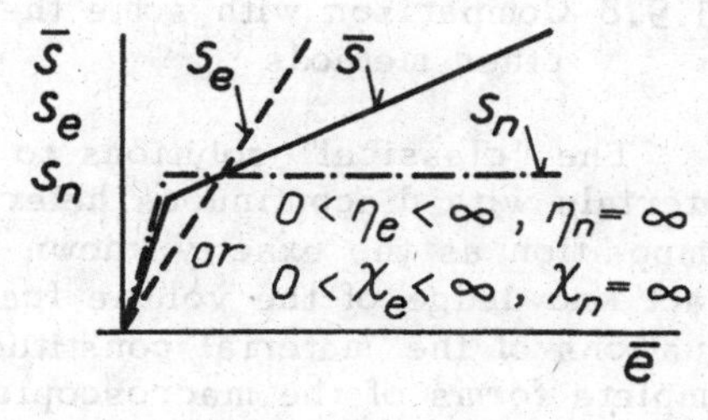

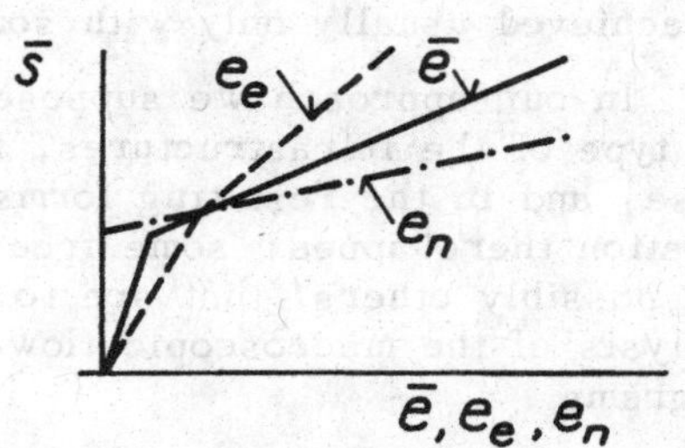

Fig. 10 Rigid-plastic inclusions in an elastic matrix

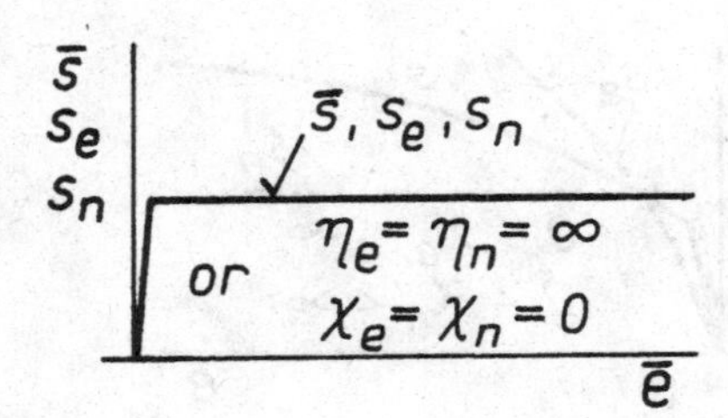

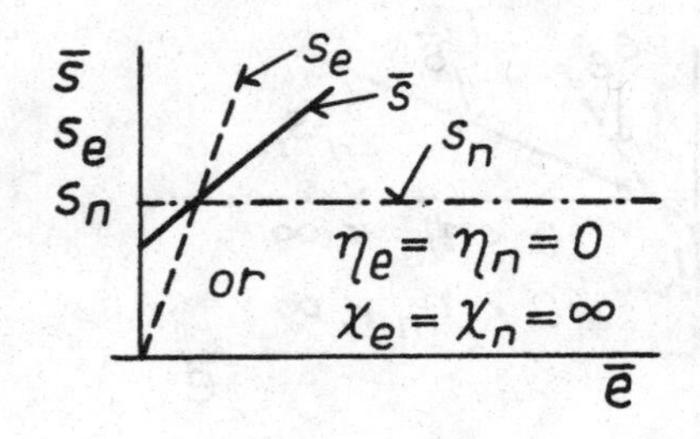

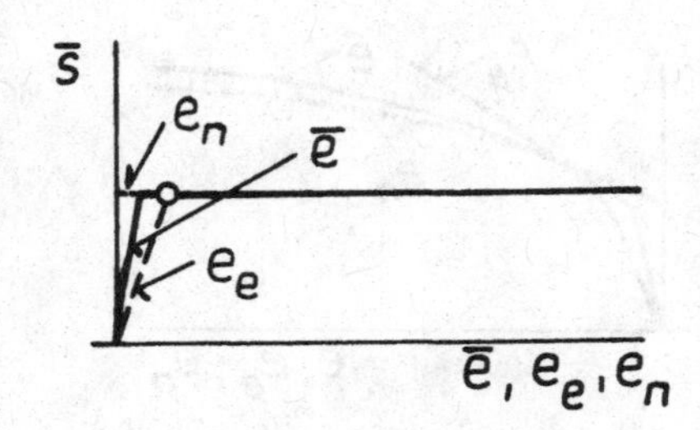

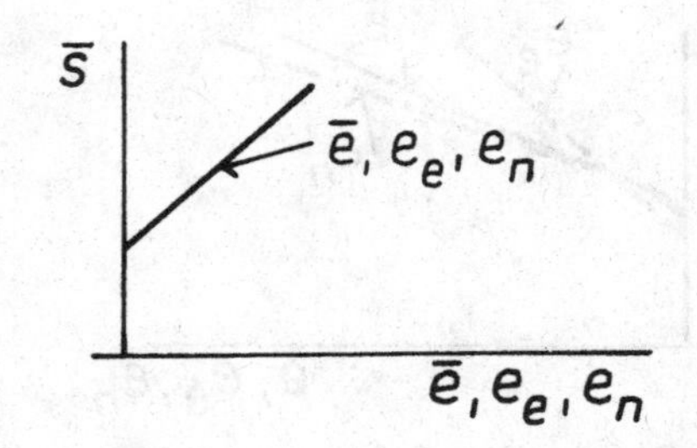

Fig. 11 Homogeneous stress - elastic and rigid-plastic constituents

Fig. 12 Homogeneous strain model - elastic and rigid-plastic constituents

I.1.9.8 Comparison with some theoretical solutions resulting from other methods

The "classical" solutions to the problems of quasihomogeneous materials with discontinuous heterogeneity consider the geometry of composition as the exactly known input information apart from the exact knowledge of the volume fractions and the constitutive equations of the material constituents. The aim is to derive the complete forms of the macroscopic constitutive equations /which can be achieved usually only with some simplifying assumptions/.

In our approach we suppose only the a priori knowledge of the type of the infrastructures, i.e. whether they are compact or loose, and in the resulting forms of the macroscopic constitutive equation there appear some free parameters /structural parameters and possibly others/ that are to be determined by a mathematical analysis of the macroscopic flow-curves or the stress-strain diagrams.

Hence, if we want to compare our models with some "classical" solutions, the exact correspondence means that it is possible to find such non-negative structural parameters η or χ , for which the respective macroscopic constitutive equations are identical. Our approach is based upon a simplifying general theorem /cf.I.1.2/ and some simplifying assumptions are generally used in the classical solutions too. Therefore, the correspondence may not be in some cases exact, but it should be at least approximate for the range that is specified by the assumptions of the classical solutions, which means very often for small volume fractions of the inclusions.

The simple fundamental solutions that are presented for comparison were taken from Reiner's monograph 63/:

a/ Rigid spherical inclusions in a viscous matrix.

Under the supposition that the volume fraction of the inclusions is very small in relation to unity A. Einstein presented the respective solution showing that the resulting macroscopic constitutive equation is of the same type as the matrix, i.e. linearly viscous, with the following macroscopic coefficient of viscosity:

$$\bar{H} = H(1 + 2.5\, v_e) \tag{107}$$

where H is the coefficient of the viscous matrix and v_e the volume fraction of the inclusions.

In our concept the respective macroscopic constitutive equation follows from eq. (97a) with $\mu_e = 0$, i.e.:

$$\dot{\bar{e}}_{ij} = \frac{v_n\, \eta_n}{2H(v_e^2 + \eta_n)}\, \bar{s}_{ij} \tag{108}$$

which means that in agreement with Einstein our resulting macroscopic constitutive equation is also linearly viscous and

$$\bar{H} = H\,\frac{v_e^2 + \eta_n}{v_n\, \eta_n} \tag{109}$$

The two expressions for $\bar{H}$ are identical for

$$\eta_n = \frac{v_e}{1.5 - 2.5\, v_e} \tag{110}$$

The structural parameter η_n is non-negative only for $v_e < 0.6$, which is not surprising with regard to the supposition under which Einstein's solution was derived.

It is interesting to note that Reiner in his monograph [63] /p.531/ states that according to his and Arnstein's experiments with mortar Einstein's equation was approximately valid up to the volume fraction of sand 0.6.

b/ Spherical bubbles in a viscous matrix.

Under a similar supposition Guth and Mark derived an analogous solution for bubbles instead of inclusions. The resulting macroscopic equation was again linearly viscous with the macroscopic coefficient of viscosity:

$$\bar{H} = H(1 - v_e) \tag{111}$$

where v_e is the volume fraction of the bubbles.

In our concept the respective macroscopic equation follows again from eq. (97a), this time with $\mu_e = \infty$, i.e.:

$$\dot{\bar{e}}_{ij} = \frac{1 + \eta_n}{2H\, v_n} \bar{s}_{ij} \tag{112}$$

which means that our macroscopic equation agrees in the type /linear viscosity/ and

$$\bar{H} = \frac{1 - v_e}{1 + \eta_n} \tag{113}$$

The two expressions for $\bar{H}$ are identical for $\eta_n = 0$.

c/ Elastic spheres in a viscous matrix.

Again under the supposition that the volume fraction of the spheres is very small compared with unity, Froehlich and Sack derived the following macroscopic constitutive equation:

$$\bar{s}_{ij} + T_1 \dot{\bar{s}}_{ij} = k(\dot{\bar{e}}_{ij} + T_2 \ddot{\bar{e}}_{ij}) \tag{114}$$

with

$$T_1 = 3H\mu_e\left(1 + \frac{5}{3} v_e\right) \tag{115}$$

$$T_2 = 3H\mu_e\left(1 - \frac{5}{2} v_e\right)$$

/there is misprint in Reiner's paper, instead of $\dot{\gamma}$ is printed γ /.

The respective equation in our concept is identical with eq. (97a), which can be rewritten in the form:

$$\bar{s}_{ij} + T_1^* \dot{\bar{s}}_{ij} = 2H \frac{v_e^2 + \eta_n}{v_n \eta_n} (\dot{\bar{e}}_{ij} + T_2^* \ddot{\bar{e}}_{ij}) \tag{116}$$

with

$$T_1^* = 2H \mu_e v_e \frac{1 + \eta_n}{v_n \eta_n} \tag{117}$$

$$T_2^* = 2H \mu_e \frac{v_e v_n}{v_e^2 + \eta_n}$$

The types of the two macroscopic equations are again identical, but for the determination of one parameter η_n we have two equations:

$$T_1 = T_1^*, \quad T_2 = T_2^*$$

From these two equations we get:

$$(\eta_n)_1 = v_e \frac{2}{3 - 5 v_e^2} \tag{118}$$

$$(\eta_n)_2 = v_e \frac{2 - 5 v_e + 7.5 v_e^2}{3 - 7.5 v_e}$$

For very small v_e we get

$$(\eta_n)_1 \doteq (\eta_n)_2 \doteq 2 v_e / 3 \tag{119}$$

for $v_e = 0.1$ the difference in the two values for η_n is about 3%.

d/ Viscous spheres in an elastic matrix.

This problem with the same supposition as in the preceding cases was solved by Oldroyd with the resulting equation:

$$\bar{s}_{ij} + T_1 \dot{\bar{s}}_{ij} = k (\bar{e}_{ij} + T_2 \dot{\bar{e}}_{ij}) \tag{120}$$

Under the assumption that the medium is incompressible it was derived for k :

$$k = \frac{1}{\mu_e} \left(1 - \frac{5}{3} v_n\right) \tag{121}$$

/there is a misprint in Reiner's paper, ν is missing /.

In our concept the respective equation is identical with eq. (98a), which can be rewritten in the form:

$$\bar{s}_{ij} + 2H \mu_e \frac{v_e \eta_e}{v_n (1 + \eta_e)} \dot{\bar{s}}_{ij} = k^* \left(\bar{e}_{ij} + 2H \mu_e \frac{v_n^2 + \eta_e}{v_e v_n} \dot{\bar{e}}_{ij} \right) \tag{122}$$

with

$$k^* = \frac{v_e}{\mu_e (1+\eta_e)}$$

Identifying k with k^* we get:

$$\eta_e = \frac{2v_n}{3-5v_n} \qquad (123)$$

which is positive for

$$v_n < 0.6$$

Thus, the types of the equations as well as the limitation for the structural parameter agree with the specification of the problem.

e/ Elastic spheres in an elastic matrix.

In this case the agreement in the type of the resulting rheological equation is of no interest - it must be elastic. Only the expressions that we get for the structural parameters are interesting.

For the inclusions we use index e, for the matrix index n.

For the bulk moduli the exact solution is given in Hashin's paper [13] : *)

$$\frac{\varrho_n}{\bar{\varrho}} = 1 + \frac{3(1-\nu_n)\left(\frac{\varrho_n}{\varrho_e} - 1\right)v_e}{2(1-2\nu_n)+(1+\nu_n)\left[\frac{\varrho_n}{\varrho_e} - \left(\frac{\varrho_n}{\varrho_e} - 1\right)v_e\right]} \qquad (124)$$

For the shear moduli, the exact solution - presented in the same paper - is known only for the case of very low volume fractions of the inclusions:

$$\frac{\mu_n}{\bar{\mu}} = 1 - \frac{15(1-\nu_n)\left(1-\frac{\mu_n}{\mu_e}\right)}{7-5\nu_n + 2(4-5\nu_n)\frac{\mu_n}{\mu_e}} v_e \qquad (125)$$

In our concept the respective equations can be arrived at from eqs. (70) and (68). We get:

$$\frac{\varrho_n}{\bar{\varrho}} = \frac{\varrho_n \overset{0}{\eta}_n + v_e(v_e \varrho_n + v_n \varrho_e)}{v_e \varrho_e + \overset{0}{\eta}_n (v_e \varrho_e + v_n \varrho_n)} \qquad (126)$$

$$\frac{\mu_n}{\bar{\mu}} = \frac{\mu_n \eta_n + v_e(v_e \mu_n + v_n \mu_e)}{v_e \mu_e + \eta_n (v_e \mu_e + v_n \mu_n)} \qquad (127)$$

***) For a special "composite spheres" assemblage**

Comparison of eqs.(124) and (125) with (126) and (127) yields:

$$\eta_n^0 = \frac{v_e(1+\nu_n)}{2(1-2\nu_n)} \tag{128}$$

$$\eta_n = \frac{2v_e(4-5\nu_n)}{7-15v_e+5\nu_n(3v_e-1)} \tag{129}$$

It is easy to see that η_n^0 is positive for the natural interval of ν_n:

$$0 \leq \nu_n \leq 0.5$$

With the same interval for ν_n, η_n is positive only for $v_e < 0.46$, which is not surprising, as eq.(127) was derived only for very small values of v_e.

From practical point of view the application of our concept to purely elastic bodies does not make much sense. The macroscopic moduli can be easily measured and it is of no use to calculate the respective structural parameters, which - on top of it - is not always possible to do with such poor input data.

However, a special importance has eq.(128), which can be used for an approximate determination of η_n^0 even in cases of inelastic deformation. Usually, inelastic deformation has only deviatoric character, and therefore, it may be difficult to determine the structural parameters for the isotropic parts, which remain elastic. Then we can use eq.(128) even if the inclusions are not exactly spherical /eq.(124) is exactly valid for any structure if the shear moduli of the constituents are equal/. Structural parameter η_n^0 can be calculated also directly from (126), but we need more information.

Concluding this paragraph we can state that in all cases an exact correspondence was found in the types of the rheological equations and it was possible to find such non-negative structural parameters that represented either an exact or a very near correspondence in the coefficients. It would be possible to proceed with other solutions, but it is of no substantial interest, the above examples, taken from Reiner's monograph , were presented only for a short demonstration.

I. 1. 10 The Solution to the Identification Problem Based on the Elastic-Plastic Stress-Strain Diagram

Here we will show the solution to the identification problem for such cases, where the elastic-plastic experimental stress-strain diagram is available and the significant feature of the process is that one of the material constituents remains elastic, whereas the other undergoes elastic-plastic deformation. The respective mathematical model will be a special case of eqs. (46) to (51) with:

$$\dot{h} = \dot{\lambda}$$
$$\dot{g}_n = \dot{T} = \dot{\tau} = 0 \tag{130}$$

The yield condition is formulated in the simplest Mises form:

$$s_{ijn}\, s_{ijn} \leq 2k^2 = \frac{3}{2} c^2 \tag{131}$$

Then our macroscopic constitutive equation will be given by eqs. (46) to (51) with $(130)_{1,2}$ and with $\dot{\lambda}$ following from (131), which gives:

$$s_{ijn}\, \dot{s}_{ijn} = 0 \tag{132}$$

For $\dot{s}_{ijn}$ we use the respective expression from $(47)_1$ and arrive at:

$$\dot{\lambda} = 0 \quad \text{for} \quad s_{ijn}\, s_{ijn} < 2k^2 \tag{133}$$

$$\dot{\lambda} = \left\langle\!\!\left\langle \frac{\mu_e(\mu_n \eta_e \eta_n + \nu_e(\mu_e \eta_e + \nu_n(\mu_n \eta_n) s_{ijn} \dot{\bar{s}}_{ij}}{\nu_e\left[(\nu_e(\mu_e \eta_e + \nu_n(\mu_n \eta_n) 2k^2 - \mu_e \eta_e\, s_{kln}\, s'_{kln}\right]} \right\rangle\!\!\right\rangle$$

$$\text{for} \qquad s_{ijn}\, s_{ijn} = 2k^2$$

where the meaning of the brackets is given as follows:

$$\langle\!\langle X \rangle\!\rangle = \frac{1}{2}\left(X + |X|\right) \tag{134}$$

In the case of an increasing uniaxial tension the deviatoric parts of stresses and strains can be expressed as products of the respective scalars and the matrix given in eq. (101). Our constitutive equation can be integrated in this case, which gives:

$$X = A(w - w_L) - B \ln \frac{1-w}{1-w_L} \tag{135}$$

$$Y = A(w - w_L) - C \ln \frac{1-w}{1-w_L} .$$

where

$$X = \frac{\bar{e} - (\bar{e})_L}{c\,\mu_e} \tag{136}$$

$$Y = \frac{\bar{s} - (\bar{s})_L}{c}$$

$$A = \frac{v_e^2\,\eta_e^2}{\eta_n (v_n + \eta_e)^2} \tag{137}$$

$$B = v_n \frac{(1+\eta_e)(\mu_n\,\eta_e\,\eta_n + v_e\,\mu_e\,\eta_e + v_n\,\mu_n\,\eta_n)}{\mu_e\,\eta_n\,(v_n + \eta_e)^2}$$

$$C = v_e\,v_n\,\frac{\mu_n\,\eta_e\,\eta_n + v_e\,\mu_e\,\eta_e + v_n\,\mu_n\,\eta_n}{\mu_e\,\eta_n\,(v_n + \eta_e)^2}$$

$$w = \frac{v_n + \eta_e}{v_e\,\eta_e} \cdot \frac{s_n'}{c} \tag{138}$$

Subscript L characterizes the values at the elastic limit.

The identification problem is solved as follows:
From eq. (135) it is easy to deduce:

$$Y' = \frac{dY}{dX} = \frac{A\,(1-w) + C}{A\,(1-w) + B} = \mu_e \frac{d\bar{s}}{d\bar{e}} \tag{139}$$

From eqs. (135) and (139) expressions w and $ln(1-w)$ are excluded and we arrive at:

$$A\,(1-w_L)(C-B) - CX + BY + (C-B)\frac{C-BY'}{1-Y'} = 0 \tag{140}$$

On a stress-strain diagram resulting from a very slow tensile test we select three suitable points on the non-linear part and determine in these points the values of X_p, Y_p, Y_p' $(p = 1, 2, 3)$ /see Appendix 4/.

We insert these three sets of values in eq. (140) and get three equations for three constants $A(1-w_L), B, C$ to be determined. These equations are not linear, but in spite of it they are easy to solve. Subtraction of two of them gives

$$B(Y_1 - Y_2 - X_1 + X_2) - (C-B)(X_1 - X_2) + \tag{141}$$

$$+ (C-B)^2 \frac{Y_1' - Y_2'}{(1-Y_1')(1-Y_2')} = 0$$

and a similar equation with indices e.g. $2,3$ instead of $1,2$. Thus, eq. (141) represents a set of two independent equations for two constants B, $(C-B)$. The first of these constants - B - can be excluded and we get a quadratic equation for $(C-B)$. It has a trivial solution $C=B$, which has no physical meaning, and a non-trivial solution:

$$C-B = (1-Y_1')(1-Y_2')(1-Y_3')\left[X_1(Y_2-Y_3) + X_2(Y_3-Y_1) + \right. \tag{142}$$

$$\left. + X_3(Y_1-Y_2)\right] \Big/ \left[(Y_1'-Y_2')(1-Y_3')(X_3-Y_3) + \right.$$

$$\left. + (Y_2'-Y_3')(1-Y_1')(X_1-Y_1) + (Y_3'-Y_1')(1-Y_2')(X_2-Y_2)\right]$$

Now, It is a straightforward matter to calculate the values of B, C, and $A(1-w_L)$ from eqs. (141) and (140).

With $A(1-w_L)$, B and C known we are able to calculate η_e, η_n and another constant, e.g. v_n. The expressions for A, B, C are given in eqs. (137), the expression for w_L follows from (138) and $(47)_{2,1}$:

$$w_L = \frac{v_n + \eta_e}{v_e \eta_e} \cdot \frac{(s_n')_L}{c} = \frac{v_n + \eta_e}{v_e \eta_e} \eta_n \frac{M'}{M} = \tag{143}$$

$$= \frac{(\mu_n - \mu_e)\eta_n (v_n + \eta_e)}{\mu_n \eta_e \eta_n + v_e \mu_e \eta_e + v_n \mu_n \eta_n}$$

After some handling a cubic equation for v_n can be derived from (137) and (143):

$$A_3 v_n^3 + A_2 v_n^2 + A_1 v_n + A_0 = 0 \tag{144}$$

where:

$$A_3 = B\mu_n(\mu_n - \mu_e) \tag{145}$$
$$A_2 = (C-B)\left[\mu_n(\mu_n - \mu_e) + B\mu_e(2\mu_n - \mu_e)\right]$$
$$A_1 = (C-B)^2\mu_e\left\{(2\mu_n - \mu_e) + \mu_e\left[B + A(1 - w_L)\right]\right\}$$
$$A_0 = (C-B)^3\mu_e^2$$

Only the solution in the interval $\langle 0,1 \rangle$ has physical meaning. With v_n and v_e found the structural parameters η_e, η_n can be determined from eqs. (137). This solves the identification problem for the case that v_n, v_e are unknown and μ_n, μ_e are known. In the case that the unknown value is either μ_n or μ_e eq. (144) can easily be transformed into a quadratic equation for μ_n or μ_e using expressions (145).

The case of inclusions of one of the material constituents in the matrix of the other one makes the analysis only easier.

The inelastic changes of volume are usually negligeable and very difficult to measure. Therefore, the structural parameters η_e^o, η_n^o related to the isotropic parts are difficult to determine. Fortunately they are not so important as the structural parameters related to the deviatoric parts changing substantially in the course of inelastic deformation. The way out is either to accept the average between the Reuss and the Voigt solutions as an approximation, or - - if the material in question is formed by a matrix with inclusions - - the two possible approaches that were discussed in I. 1.9.8. e. If the bulk moduli of the two material constituents do not differ substantially, the field of isotropic parts of stress- and strain- tensors can be described approximately as homogeneous.

I. 1. 11 Applications to Polycrystalline Metals

It is one of the main ideas of our concept that the scale, on which an operative model should be based, is the one-step-down scale from macro- to micro- . The application of this point of view to polycrystalline metals is not quite straightforward. There are two possibilities: the first one is to describe a polycrystalline material as an assemblage of anisotropic grains with different orientations, the other one is to consider the difference between the inner regions of grains and their boundary regions as the main feature and basis for description. In our concept the last approach was accepted and applied. It is believed that it represents the most important mechanism for technical materials and alloys, where the boundary regions contain different kinds of impurities and imperfections. The anisotropic character of individual grains is not included in the analysis, only their averaged isotropic behaviour is considered.

Such a point of view is corroborated by the fact, that the common technically important materials are composed of crystals with a sufficient number of slip-planes that enable any inelastic deviatoric deformation of the crystal as a whole without creating any important and systematic rise of elastic energy due to the anisotropy of their resistance.

On the other hand the difference between the resistance of the inner regions and the boundary regions of grains creates important and systematic microstress-fields, as was clearly shown by the X-ray diffraction method by D.M. Vasilyev and his colleagues [74,75,76]. /The boundary regions do not play any important role in very pure metals at the beginning of plastic deformation, but gradually dislocations pile up at the boundaries and the resulting imperfections in the crystallographic structure represent an important factor even in this case/.

In waht follows it will be shown that a mathematical model based on this point of view is substantiated, as it is able to describe a number of important phenomena.

I.1.11.1 Determination of the Volume Fraction of the Material Constituents

According to the preceding paragraph it is possible to determine - among others - the volume fractions by a mathematical analysis of the stress-strain curve and - if the volume fractions are known - to check the possibilities of the model.

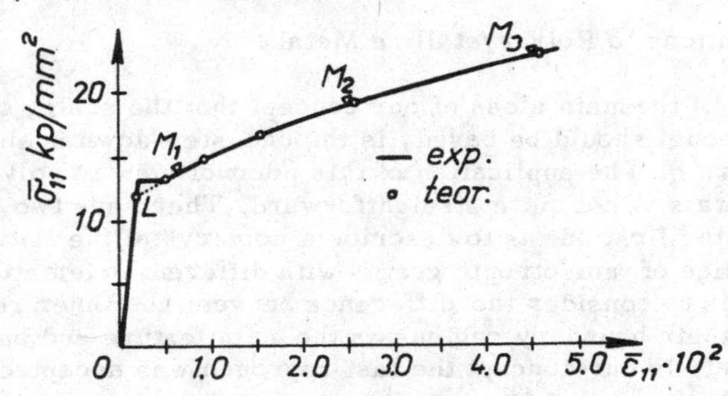

Fig. 13 Stress-strain diagram of an aluminium alloy /after W. Szczepiński [68]/.

For a numerical example we will use the stress-strain diagram /Fig. 13/ of an aluminium-alloy specimen under quasistatic very slowly increasing uniaxial tension that was published by W. Szczepiński [68]. The material in question was composed of 94.05% Al and 5.95% ingredients of Mg, Mn, Zn, Si, Fe, Cr.

To simplify the analysis we will consider the material as homogeneous in regard to its elastic properties and suppose that in the course of plastic deformation some continuous subset of microvolumes remains elastic, whereas the remainder of the material deforms plastically.

This means that

$$\mu_n = \mu_e = \bar{\mu} = \mu \ , \quad \varrho_e = \varrho_n = \bar{\varrho} = \varrho$$
$$w_L = 0$$

and eqs. (133), (135), (136), (139), (140), (143), (144), (145) will change accordingly.

Young's modulus was determined from the slope of the straight elastic part of the diagram, the elastic limit from the dotted backward extrapolation /cf. Appendix 4./:

$$E = 6640 \ kp/mm^2$$

$$c = (s_{11n})_L = (\bar{s}_{11})_L = \frac{2}{3}(\bar{\sigma}_{11})_L = 7.95 \ kp/mm^2$$

and with Poisson's ratio

$$\nu = 0.3$$

we get:

$$\mu = (1+\nu)/E = 1.96 \cdot 10^{-4} \ mm^2/kp$$

In the three selected points P_1, P_2, P_3 the values of X_p, Y_p, Y'_p $(p = 1, 2, 3)$ were determined as follows :

p	X_p	Y_p	Y'_p
1	2.9897	0.162	0.0486
2	14.9153	0.587	0.02895
3	27.5867	0.912	0.02153

Solution to eqs. (142), (141) and (140) gives:

$$A = 0.36777$$

$$B = 8.91788$$

$$C = 0.18169$$

and solution to eqs. (144), (137):

$$v_e = 1 - v_n = 0.06362$$

$$\eta_e = 2.12287$$

$$\eta_n = 5.3003 \cdot 10^{-3}$$

The agreement between the calculated and the a priori known volume fraction /6,36% and 5,95%/ is very good. In another paper /Kafka [26]/ / the same experimental stress-strain curve with the same three selected points was analysed using another form of the yield condition that included also the influence of the s'_{ijn} deviator. The calculated volume fraction was 5,93%, i.e. an even better agreement with the real value.

With the values v_e, η_e, η_n determined it is easy to plot the theoretical stress-strain diagram as is shown in Fig. 13.

In the paper by Kafka and Štětina [34]/ a similar calculation was performed for the eutectic alloy AlSi 12 where the known volume fraction of Si was 14,2%. The volume fraction was calculated from stress-strain diagrams of three specimens. The average calculated value was 14,8%, the greatest deviation was 4.7%.

In his MSc thesis J. Jirků [23]/ calculated the volume fraction of Si in AlSi 13 Mn alloy with the assumption of different μ_n and μ_e values, as they are known from homogeneous samples experiments. The calculated value was 12.73%, the a priori known value 14.76%.

A higher difference in the two last named cases is probably due to the fact that here the experiments were performed only in common technical conditions - deformations measured from the displacements of the platens of the machine.

I.1.11.2 Determination of the Stored Energy after Plastic Deformation

Using the same solution to the identification problem as in the preceding section we can evaluate the energy that is stored in the material after a macroscopically homogeneous plastic deformation process.

A detailed explanation is advanced in Kafka /33,29/. The stored energy is expressed by the energy of residual deviatoric stresses. The effect of the isotropic parts of microstresses is neglected and so is the loss of residual energy due to the relaxation of microstresses.

The process of unloading is supposed to be elastic and thus, the values of the residual deviatoric microstresses are given by the formulae:

$$s^r_{ijn} = s_{ijn} - (s_{ijn})_{el} \;, \qquad s^r_{ije} = s_{ije} - (s_{ije})_{el} \tag{146}$$

$$s'^r_{ijn} = s'_{ijn} - (s'_{ijn})_{el} \;, \qquad s'^r_{ije} = s'_{ije} - (s'_{ije})_{el}$$

and according to eq. (18) we arrive at the following formula for the deviatoric stored energy:

$$\overset{r}{W}_d = \frac{1}{2} v_e \mu_e \left[s^r_{ije} s^r_{ije} + \frac{1}{\eta_e} s'^r_{ije} s'^r_{ije} \right] + \tag{147}$$

$$+ \frac{1}{2} v_n \mu_n \left[s^r_{ijn} s^r_{ijn} + \frac{1}{\eta_n} s'^r_{ijn} s'^r_{ijn} \right]$$

In the case of homogeneous elastic properties ($\mu_n = \mu_e$) we get:

$$s^r_{ijn} = s_{ijn} - \bar{s}_{ij} \;, \qquad s^r_{ije} = s_{ije} - \bar{s}_{ij} \tag{148}$$

$$s'^r_{ijn} = s'_{ijn} \;, \qquad s'^r_{ije} = s'_{ije}$$

The value $\overset{r}{W}_d$ can easily be calculated for the current point of the stress-strain diagram and the course of stored energy plotted and compared with the experimental values arrived at by calorimetric measurements. For one of the materials that were studied in Kafka /33/ the result is shown in Fig. 14.

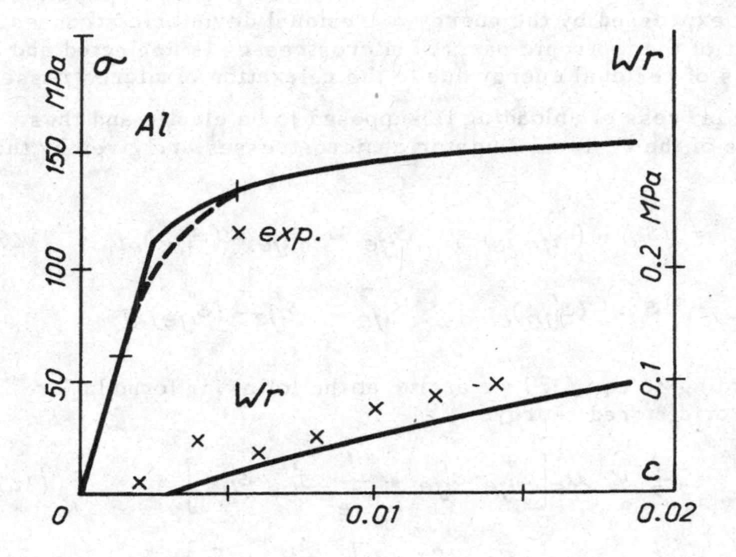

Fig. 14 Calculated and measured course of stored energy /after V. Kafka [33]/.

Apart from the agreement with calorimetric measurements another important agreement must be emphasized. With the X-ray diffraction method the second-order internal stresses can well be determined and it is observed /cf. Titchener and Bever [69] and Bolshanina and Panin [4]/ that the energy represented by these microstresses is only about 10% of the total stored energy /as measured by the calorimetric method/. This is in an excellent accord with our calculations, as the energy resulting from the expression

$$\frac{1}{2} v_e \mu_e s^r_{ije} s^r_{ije} + \frac{1}{2} v_n \mu_n s^r_{ijn} s^r_{ijn}$$

is about 10% of the total value, which means that about 90% of the stored energy are due to the effect of fluctuations represented by the formula:

$$\frac{v_e \mu_e}{2 \eta_e} s'^r_{ije} s'^r_{ije} + \frac{v_n \mu_n}{2 \eta_n} s'^r_{ijn} s'^r_{ijn}$$

Hence, our approach opens a possibility of calculating stored energy and the comparisons with the results of calorimetric measurements mean another verification of the adequacy of our model for application to polycrystalline materials.

I.1.11.3 Changes of the Yield Surfaces

In the preceding sections the yield conditions were considered in the simplest Mises form, i.e. for the inelastic constituent it was assumed:

$$s_{ijn}\, s_{ijn} \leq 2k^2 \tag{149}$$

and for the solution of a number of problems this simple assumption turns out to be sufficient.

If the elastic properties of the material constituents are homogeneous, then it holds according to $(49)_1$ and $(47)_1$ in the elastic range:

$$M = 1 \quad , \qquad s_{ijn} = \bar{s}_{ij} \tag{150}$$

and the combination of the above equations gives the macroscopic yield condition in the virgin state:

$$\bar{s}_{ij}\, \bar{s}_{ij} = 2k^2 \tag{151}$$

i.e. the Mises criterion on the macroscale.

After some plastic deformation and unloading there remain in the material constituents residual microstresses and among them the residual microstress component s^r_{ijn} . For a new loading we have

$$s_{ijn} = s^r_{ijn} + \bar{s}_{ij} \tag{152}$$

and the combination of eqs. (149) and (152) gives the new macroscopic yield condition:

$$(s^r_{ijn} + \bar{s}_{ij})(s^r_{ijn} + \bar{s}_{ij}) \leq 2k^2 \tag{153}$$

i.e. a translation of (151).

A detailed study of this problem led to the conclusion that for many materials the changes of the macroscopic yield surface depend not only on the components s^r_{ijn} , but also on s'^r_{ijn} .

This influence can be included into the mathematical analysis as follows:

- The direction in a space of symmetric deviators can be defined by a unit deviator S_{ij} with the properties:

$$S_{ij}\, S_{ij} = 1 \, , \quad S_{ii} = 0 \tag{154}$$

- Two directions S_{ij} and $S^{\perp}_{ij}$ with the property:

$$S_{ij}\, S^{\perp}_{ij} = 0$$

are called orthogonal.

Thus, e.g. uniaxial tension in the x_1 - direction or in the x_2 - direction or pure shear in the $x_1 . x_2$ directions have their deviatoric directions respectively:

$$S_{ij}^{(1)} = \begin{pmatrix} \sqrt{\frac{2}{3}} & 0 & 0 \\ 0 & -\sqrt{\frac{1}{6}} & 0 \\ 0 & 0 & -\sqrt{\frac{1}{6}} \end{pmatrix}, \quad S_{ij}^{(2)} = \begin{pmatrix} -\sqrt{\frac{1}{6}} & 0 & 0 \\ 0 & \sqrt{\frac{2}{3}} & 0 \\ 0 & 0 & -\sqrt{\frac{1}{6}} \end{pmatrix}, \quad S_{ij}^{(12)} = \begin{pmatrix} 0 & \sqrt{\frac{1}{2}} & 0 \\ \sqrt{\frac{1}{2}} & 0 & 0 \\ 0 & 0 & 0 \end{pmatrix}$$

It is clear enough that:

$$S_{ij}^{(1)} S_{ij}^{(12)} = S_{ij}^{(2)} S_{ij}^{(12)} = 0$$

$$S_{ij}^{(1)} S_{ij}^{(2)} \neq 0$$

which means that uniaxial tensions in the x_1 - direction and the x_2 - direction have not orthogonal deviatoric directions in our convention, but any of the two directions of tension is orthogonal to the direction of the pure shear.

- The scalar quantity $S_{ij} T_{ij}$ will be called the component of the deviator T_{ij} in the direction S_{ij}.

- Any deviator T_{ij} of the field has its direction S_{ij}^T defined as follows:

$$S_{ij}^T = \frac{T_{ij}}{\sqrt{T_{kl} T_{kl}}} \tag{155}$$

- It holds:

$$S_{ij}^T T_{ij} > S_{ij} T_{ij} \quad \text{for} \quad S_{ij} \neq S_{ij}^T \tag{156}$$

<u>Proof:</u>

Let us define: $\quad Q_{ij} = S_{ij}^T - S_{ij}$

$$1 = S_{ij}^T S_{ij}^T = (S_{ij} + Q_{ij})(S_{ij} + Q_{ij}) = S_{ij} S_{ij} + Q_{ij} Q_{ij} +$$

$$+ 2 S_{ij} Q_{ij} = 1 + Q_{ij} Q_{ij} + 2 S_{ij} Q_{ij} .$$

from which:

$$S_{ij}\,Q_{ij} = -\frac{1}{2}\,Q_{ij}\,Q_{ij} < 0$$

$$S_{ij}\,S_{ij}^{T} = S_{ij}\,(S_{ij} + Q_{ij}) = 1 + S_{ij}\,Q_{ij} < 1$$

$$S_{ij}\,S_{ij}^{T} = S_{ij}\,\frac{T_{ij}}{\sqrt{T_{kl}\,T_{kl}}} < 1 = S_{ij}^{T}\,S_{ij}^{T} = S_{ij}^{T}\,\frac{T_{ij}}{\sqrt{T_{kl}\,T_{kl}}}$$

which gives:

$$S_{ij}\,T_{ij} < S_{ij}^{T}\,T_{ij}$$

q. e. d.

/End of the proof./

Furthermore. according to eq. (18) the specific elastic energy appertaining to one deviatoric component /including the influence of fluctuations/ in the n-th material constituent is expressed as

$$\frac{1}{2}\mu\left[(s_{\alpha\beta n})^{2} + \frac{1}{\eta_{n}}(s'_{\alpha\beta n})^{2}\right]$$

The specific elastic energy appertaining to the component in the direction S_{ij} will be expressed analogously:

$$\frac{1}{2}\mu\left[(S_{ij}\,s_{ijn})^{2} + \frac{1}{\eta_{n}}(S_{ij}\,s'_{ijn})^{2}\right]$$

The yield condition is newly formulated in the following way: "The specific elastic energy of a deviatoric stress component corresponding to any direction cannot overpass a certain limit."

In mathematical terms this yield criterion will read:

$$(S_{ij}\,s_{ijn})^{2} + \frac{1}{\eta_{n}}(S_{ij}\,s'_{ijn})^{2} \leq 2k^{2} \tag{157}$$

In the elastic range of a virgin material with homogeneous elastic properties $s'_{ijn} = 0$ and $S_{ij}\,s_{ijn}$ will achieve the greatest value for

$$S_{ij} = \frac{s_{ijn}}{\sqrt{s_{kln}\,s_{kln}}} \tag{158}$$

In such a case eq. (157) transforms to:

$$s_{ijn}\, s_{ijn} \leq 2k^2 \tag{159}$$

i.e. Mises' criterion, a hypersphere, an isotropic condition.

In the course of an uniaxial plastic deformation in the direction

$$\mathcal{S}^1_{ij} = \frac{\bar{s}_{ij}}{\sqrt{\bar{s}_{kl}\, \bar{s}_{kl}}} \tag{160}$$

the respective microstresses s_{ijn}, s'_{ijn} will have - according to eqs. $(47)_1$, $(47)_2$ - the same direction:

$$\mathcal{S}^1_{ij} = \frac{s_{ijn}}{\sqrt{s_{kln}\, s_{kln}}} = \frac{s'_{ijn}}{\sqrt{s'_{kln}\, s'_{kln}}} \tag{161}$$

and the current yield surface results from (161) and (157):

$$s_{ijn}\, s_{ijn} + \frac{1}{\eta_n}\, s'_{ijn}\, s'_{ijn} = 2k^2 \tag{162}$$

or with the use of eq. (152):

$$(\bar{s}_{ij} + s^r_{ijn})(\bar{s}_{ij} + s^r_{ijn}) = 2k^2 - \frac{1}{\eta_n}\, s'_{ijn}\, s'_{ijn}$$

where - according to (152):

$$\frac{s^r_{ijn}}{\sqrt{s^r_{kln}\, s^r_{kln}}} = -\mathcal{S}^1_{ij} \tag{163}$$

This means that <u>the yield surface is shifted in the direction of the acting loading and it gets narrower in this direction.</u>

If the uniaxial loading process in the $\mathcal{S}^1_{ij}$ direction stops and the macrostress will change inside the yield surface, i.e. without further plastic deformation, the stress components s'_{ijn} and s^r_{ijn} will remain unchanged, but s_{ijn} will change in accordance with eq. (152). Let us suppose that the macrostress will be changed to reach the value:

$$\bar{s}^I_{ij} = -s^r_{ijn} \tag{164}$$

In this state we have from (152):

$$s_{ijn} = 0$$

Furthermore, let $\bar{s}^{I}_{ij}$ be held fixed and a new loading process in a perpendicular direction $\overset{2}{S}_{ij}$ be started; i.e.:

$$\overset{2}{S}_{ij}\,\overset{1}{S}_{ij} = 0 \tag{165}$$

$$\overset{2}{S}_{ij} = \frac{\bar{s}_{ij} - \bar{s}^{I}_{ij}}{\sqrt{(\bar{s}_{kl} - \bar{s}^{I}_{kl})(\bar{s}_{kl} - \bar{s}^{I}_{kl})}} = \frac{\bar{s}_{ij} + s^{r}_{ijn}}{\sqrt{(\bar{s}_{kl} + s^{r}_{kln})(\bar{s}_{kl} + s^{r}_{kln})}} =$$

$$= \frac{s_{ijn}}{\sqrt{s_{kln}\, s_{kln}}} \tag{166}$$

For this new perpendicular direction $\overset{2}{S}_{ij}$ the yield condition will be:

$$(s_{ijn}\,\overset{2}{S}_{ij})^2 + \frac{1}{\eta_n}(s'_{ijn}\,\overset{2}{S}_{ij})^2 = 2k^2 \tag{167}$$

The value of s'_{ijn} remained unchanged and therefore /cf. (161)/:

$$s'_{ijn}\,\overset{2}{S}_{ij} = \sqrt{s'_{kln}\, s'_{kln}}\;\overset{1}{S}_{ij}\,\overset{2}{S}_{ij} = 0 \tag{168}$$

and (167) with (166) gives:

$$s_{ijn}\, s_{ijn} = (\bar{s}_{ij} - \bar{s}^{I}_{ij})(\bar{s}_{ij} - \bar{s}^{I}_{ij}) = 2k^2 \tag{169}$$

which means that the width of the yield surface in the perpendicular direction remained unchanged.

These fundamental characteristics of the changes of the yield surfaces agree with experimental evidence as published e.g. by A. Phillips et al. /56,58/

I. 1. 11.4 The Influence of the Rate of Loading upon the Stress-Strain Curve and upon the Strength

It is a well known property of the metallic as well as other materials that the inelastic part of the stress-strain curve is steeper at a higher rate of loading. This phenomenon can be described by our model supposing that there exist two infrastructures in the material: one with elastic properties and the other with elastoviscoplastic properties.

The model that we use for the elastoviscoplastic properties is a special case of that described in section I.1.9. Here, expression (53) takes on the form:

$$\dot{h} = \dot{\beta}$$

where

$$\dot{\beta} = 0 \qquad \text{for} \qquad s_{ijn}\, s_{ijn} < 2\varkappa^2$$

$$\dot{\beta} = \frac{1}{2H} \cdot \frac{\sqrt{s_{ijn}\, s_{ijn}} - \sqrt{2}\,\varkappa}{\sqrt{s_{ijn}\, s_{ijn}}} \qquad \text{for} \quad s_{ijn} s_{ijn} > 2\varkappa^2$$

It can easily be shown that this model corresponds to the so called Bingham body. Thus, e.g. for

$$s_{12n} = s_{21n} \neq 0 \;, \quad s_{ijn} = 0 \qquad \text{for} \quad ij \neq 12\,,21$$

equation $(44)_1$ takes on the form:

$$\dot{e}_{12n} = \mu_n\, \dot{s}_{12n} + \frac{1}{2H}\left(s_{12n} - \varkappa\right)$$

which is the well known Bingham equation.

There are three possibilities:

If the loading rate $\dot{\bar{s}}_{ij}$ is very high, the second addend on the right-hand side of eq. $(47)_1$ is negligible and the process is purely elastic. If $\dot{\bar{s}}_{ij}$ is very low, the stress-strain curve corresponds to a sequence of states with ended viscous flow, i.e. to an elastic-plastic process. For a finite rate of loading the process is elastoviscoplastic, depending on the rate.

The stress-strain curves following from these equations are plotted in Fig. 15.

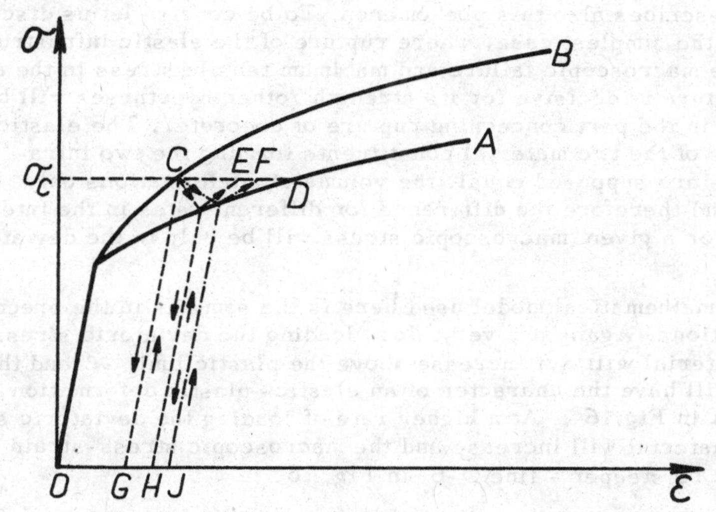

Fig. 15 . Stress-strain curves at different rates of loading.

At a very slow /quasistatic/ loading the process is merely elastic-plastic. There is enough time for the deviatoric stress in the n -constituent to be reduced at any stage to the value $\mathcal{H}$. under which there is no more flow /line O-A in Figs. 15 and 16/.

At a higher rate of loading the values of R will depend on the given rate of loading, the value of the deviatoric stress in the n- constituent will not be constant as in the preceding case, it will increase. and the macroscopic stress-strain curve will be steeper /line O-B in Figs. 15 and 16/.

If the increase of loading proceeding at a finite rate is stopped in point C /Fig. 15/, the inelastic strain will increase and after some time /theoretically infinite/ it will reach point D . If the reaching of point C is followed by a very quick unloading and reloading to point C and then by the increase of loading with the original finite rate, it will correspond to the course O-C-G-C-B, i.e. to a scheme that is similar to the time- independent plasticity. A slow unloading from point C and slow reloading up to the level of σ_C leads to the trace O-C-H-E-D. A still slower unloading and reloading to the trace O-C-J-F-D. These characteristic features of the mathematical model fully agree with experimental findings as published by A. Phillips et al. [56,58] /

Another generally known property is the higher strength appearing at higher rate of loading. Our mathematical model explains and quantitatively describes also this phenomenon. To be concise let us discuss here only the simplest case, where rupture of the elastic infrastructure causes the macroscopic failure and maximum tensile stress in the elastic infrastructure is decisive for its strength /other hypotheses will be discussed in the part concerning rupture of concrete/. The elastic properties of the two material constituents forming the two infrastructures are supposed equal, the volumetric deformations only elastic, and therefore the difference for different rates in the internal stresses for a given macroscopic stress will be only in the deviatoric parts.

The mathematical model used here is the same as in the preceding considerations. Again at a very slow loading the deviatoric stress in the n - material will not increase above the plastic limit $\varkappa$ and the process will have the character of an elastic- plastic deformation - - line O-A in Fig.16 . At a higher rate of loading the deviatoric stress in the n-material will increase and the macroscopic stress-strain curve will be steeper - line O-B in Fig. 16.

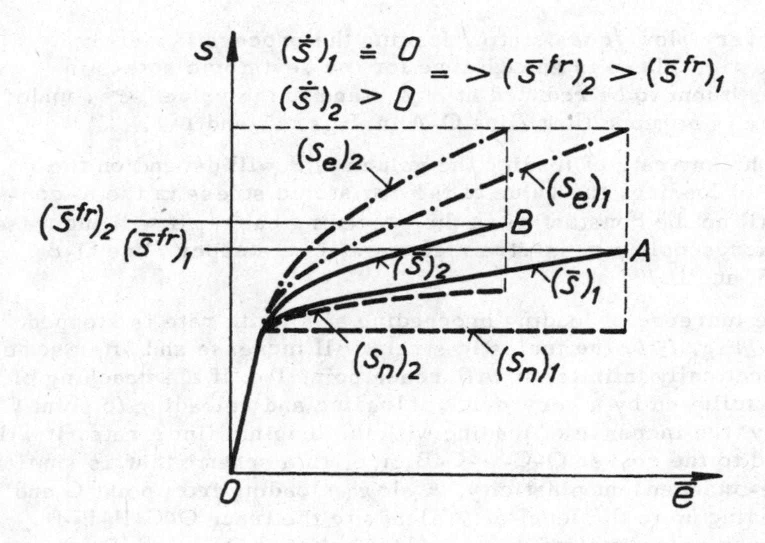

Fig. 16. Strength at different rates of loading

The respective courses of deviatoric stress in the e-constituent and the n-constituent, as plotted in Fig. 16, show that a certain limit in the e-material is reached at a higher macroscopic stress and lower macroscopic deformation if the rate of loading is higher. The isotropic component of stress in the e-material does not depend on the rate of loading and therefore the tensile stress in the e-constituent and - according to our hypothesis - the resulting macroscopic failure is to appear at a higher macroscopic stress and a lower macroscopic deformation if the rate of loading is higher. This agrees with the experimental evidence.

I. 1. 11.5 The yield-point jog

The phenomenon of yield-point jog is often explained (cf. e. g. J. Němec 53/, p. 195) by the resistance of boundary regions of grains with their concentrated impurities. Up to the yield-point jog they form a system of shells that close inside the "inclusions" of the inner parts of grains, where plastic deformation can easily proceede. The yield-point jog means cracking of this system of shells.

In our concept this scheme can well be described by two infrastructures: that of an elastic material corresponding to the system of shells and that of an elastic-plastic material corresponding to the inner parts of grains. Up to the yield-point jog the infrastructure of the elastic-plastic inner parts is loose, it forms "inclusions", and therefore the respective structural parameters are infinite. At the yield-point jog the separating system of shells is broken, which forms bridges for plastic deformation. In this way the infrastructure of the elastic-plastic material constituent becomes continuous, or - in our terminology - compact, with finite structural parameters. The infrastructure of the elastic constituent is supposed to be continuous before as well as after its cracking at the yield-point jog, but the respective finite structural parameters will be different. /It does not seem probable that cracking at the yield-point jog would completely disintegrate the system of shells into separated inclusions./

In Fig. 17 the courses of the macroscopic deviatoric stress and the average stresses in the two material constituents were plotted by the HP20 calculator on the basis of our mathematical model with randomly chosen parameters.

Let us mention that this mathematical model of the yield-point jog effect agrees with experimental evidence also in these points:

a/The slope of the stress-strain diagram is slightly curved before reaching the yield-point jog itself.

b/ The average stress in the elastic infrastructure /shells/ drops due to the yield-point jog, nevertheless it remains higher than that in the plastic material constituent /inner parts of grains/. This agrees with the experimental findings /X-ray diffraction method/ by D.M. Vasilev and L.V. Kozevnikova [76/].

c/ A very high concentration of acoustic emission signals is observed at the yield-point jogs - communicated e. g. by A. Peter, A. Fehervary [55/].

From the point of view of the identification problem it results that the structural parameters corresponding to the elastic part of the stress-strain curve can differ from those corresponding to the elastic-plastic part if there appears a yield-point jog.

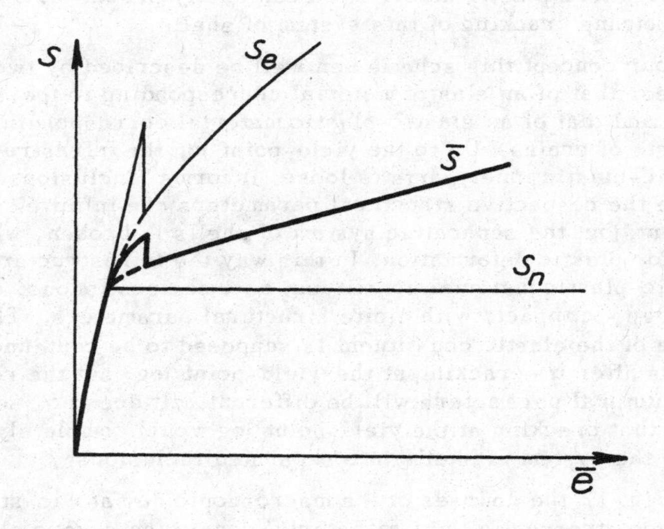

Fig. 17 The yield-point jog and the respective microstresses

I.1.11.6 The influence of the loading history upon the stress-strain diagram and upon the strength

The effect of the loading history has two different features. One of them is the fading memory effect, represented by reversible changes, the other one is the effect of irreversible changes that - if proceeding long enough - leads to rupture.

In our concept the reversible effect is represented by residual deviatoric microstresses that remain in the material after plastic or viscous deformation process. It is clear enough from eqs. (47) and (1) that inelastic loading process and elastic unloading lead to residual stresses in the material constituents that form a self-equilibrated field of internal stresses.

In what follows we will confine ourselves to the case of homogeneous elastic properties. Then in the course of an uniaxial loading the average deviatoric stress in the inelastic /elastic/ material constituent will be lower /higher/ than the macroscopic stress and elastic unloading results therefore in residual stresses that are in the inelastic /elastic/ material constituent of an opposite /equal/ sign as the loading stress. In the case of a plastic deformation process the residual stresses will remain unchanged until a new loading steps in. If this new loading has the same direction as the first one, the response will be elastic up to the value of the macroscopic stress where the unloading in the first process began. In this point the yield surface is reached and the stress-strain curve continues as without the elastic unloading and reloading. If the new loading has another direction than the first one, a different pattern of residual stresses will be formed and the deviatoric residual stresses from the first process will be gradually "erased".

In the case of a viscous material the residual stresses after unloading will relax and the fading of memory will appear even without any further loading.

The irreversible effect is represented by residual isotropic microstresses that are always compressive in the inelastic material constituent /owing to the violation of the regularity of the atomic lattice by plastic deformation that results in the increase of volume and due to the constraint by the surrounding medium that resists this increase of volume/ and from eq. (1) with zero macroscopic stress we can easily deduce /multiplying this equation by Kronecker's δ / that the respective residual isotropic microstresses in the elastic material constituent must be of an opposite sign, i.e. tensile.
Whatever be the subsequent loading process, these isotropic components are not erased, they can only increase.

The above mentioned signs and changes of the residual deviatoric and isotropic parts of the average stresses in the two material constituents follow from our mathematical model, but there is also available experimental evidence to support this concept. D.M. Vasilev [74,75,76] established /X-ray diffraction method/ for all tested materials without exception /several kinds of steel, iron, aluminium, nickel, copper and molybdenum/ the following inequalities for residual stresses:

$$\sigma_{11n}^{r} < \sigma_{22n}^{r} < 0 , \quad \sigma_{11e}^{r} > \sigma_{22e}^{r} > 0 \qquad \text{for tension} \quad \bar{\sigma}_{11} > 0$$

$$\sigma_{22n}^{r} < \sigma_{11n}^{r} < 0 , \quad \sigma_{22e}^{r} > \sigma_{11e}^{r} > 0 \qquad \text{for compression} \quad \bar{\sigma}_{11} < 0$$

with

$$\sigma_{11n}^{r} = s_{11n}^{r} + \sigma_{n}^{r} , \quad \sigma_{11e}^{r} = s_{11e}^{r} + \sigma_{e}^{r}$$

and

$$s_{11n}^{r} + s_{22n}^{r} + s_{33n}^{r} = s_{11n}^{r} + 2\, s_{22n}^{r} = 0$$

$$s_{11e}^{r} + 2 s_{22e}^{r} = 0$$

it is easy to deduce from the above inequalities:

$$s_{11n}^{r} < 0 , \quad s_{11e}^{r} > 0 \qquad \text{for tension}$$

$$s_{11n}^{r} > 0 , \quad s_{11e}^{r} < 0 \qquad \text{for compression}$$

$$\sigma_{n}^{r} < 0 , \quad \sigma_{e}^{r} > 0 \qquad \text{for tension as well as for compression}$$

which agrees with the consequences of our theoretical concept.

From the macroscopic point of view microstresses can be considered as latent variables descriptive of the state of the material in plastic or viscous process /cf. V. Kafka [27]/. T. Inoue and K. Yamamoto [21] used our concept for the description of the plastic deformation response of an aluminium material to complicated loading paths with a very good result. Nevertheless the formulation of the best suitable yield condition for the material in question in terms of s_{ijn} and s_{ijn}' is a problem that calls for further investigation.

Finally let us shortly hint at the agreement of our concept with the so called "Manson - Coffin equation". As mentioned above, in our concept the measure of exhaustion of strength in the case of plastic deformation is represented by the irreversible increase of the tensile isotropic component σ_e^r in the e-material constituent, i.e.:

$$\sigma_e^r \leq (\sigma_e^r)_{crit} = K_c \tag{170}$$

or

$$K_c = \int_{\sigma_e^r = 0}^{\sigma_e^r = K_c} d\sigma_e^r \tag{171}$$

According to eq. (1) it holds:

$$d\sigma_e^r = -\frac{\nu_n}{\nu_e} d\sigma_n^r$$

and eq. $(47)_3$ gives:

$$d\sigma_n^r = -N_0 \, d\tau$$

/we do not describe any quasihomogeneous fracturing or temperature changes in this case and therefore $\dot{T} = \dot{\varrho}_n = 0$ /. The increase of volume in the n-material constituent is supposed to be a consequence of plastic deformation. Let us describe this relation in the simplest way possible:

$$d\tau = K_\tau \, d\lambda \tag{174}$$

Furthermore, the increment of macroscopic plastic deformation follows from eqs. (46) and (53) /here $M = 1$, $M' = 0$ due to the assumed elastic homogeneity/:

$$d\bar{\varepsilon}_{11}^{pl} = \nu_n \, s_{11n} \, d\lambda \tag{175}$$

In the case that the simplest Mises condition /eq. (103) / is used for the average deviatoric stress s_{ijn} and the loading in question is uniaxial stress $\bar{\sigma}_{11}$, it holds throughout the plastic process:

$$|s_{11n}| = c \tag{176}$$

The set of the above equations results in:

$$K_c = \int_{\sigma_e^r = 0}^{\sigma_e^r = K_c} \frac{N_0 K_\tau}{\nu_e c} \left| d\bar{\varepsilon}_{11}^{pl} \right| \doteq \frac{N_0 K_\tau}{\nu_e c} \sum_{\sigma_e^r = 0}^{\sigma_e^r = K_c} \left| \Delta \bar{\varepsilon}_{11}^{pl} \right| \tag{177}$$

Supposing that in any cycle the increment $|\Delta \bar{\varepsilon}_{11}^{pl}|$ is the same, we can rewrite the last equation as follows:

$$N_\Delta \left| \Delta \bar{\varepsilon}_{11}^{pl} \right| = K \tag{178}$$

where

$$K = \frac{K_c \, v_e \, c}{N_0 \, K_\tau} \tag{179}$$

and N_Δ is the number of cycles up to rupture for a given value of $|\Delta \bar{\varepsilon}_{11}^{pl}|$.

Hence, we have arrived at a special case of the Manson-Coffin equation

$$N_\Delta^{\beta} \, |\Delta \bar{\varepsilon}_{11}^{pl}| = K \tag{180}$$

for $\beta = 1$, which corresponds to processes at higher temperatures. At lower temperatures $\beta < 1$.

It is natural to assume that at higher temperatures the fluctuations of the microstress field in the n-material constituent do not play any important role as they are relaxed by creep. Therefore criterion (103) and the corresponding equation (176) are acceptable. On the other hand at lower temperatures criterion (157) that takes into account the heterogeneity of the microstress field may better describe the real behaviour. For our uniaxial loading $\bar{\sigma}_{11}$ this criterion takes on the form:

$$s_{11n}^2 + \frac{1}{\eta_n} s_{11n}^{\prime 2} = c^2$$

or

$$s_{11n} = \sqrt{c^2 - \frac{1}{\eta_n} s_{11n}^{\prime 2}} \tag{181}$$

With $\beta < 1$ equation (180) leads to:

$$N_\Delta = 1 \quad => \quad N_\Delta |\Delta \bar{\varepsilon}_{11}^{pl}| = |\bar{\varepsilon}_{11}^{pl}|_{crit.} = K \tag{182}$$

$$N_\Delta > 1 \quad => \quad N_\Delta |\Delta \bar{\varepsilon}_{11}^{pl}| = |\bar{\varepsilon}_{11}^{pl}|_{crit.} > K \tag{183}$$

Similar relations result from our concept: Using (181) instead of (176), we have:

$$K_c = \int_{\sigma_e^r = 0}^{\sigma_e^r = K_c} \frac{N_0 \, K_\tau}{v_e \sqrt{c^2 - \frac{1}{\eta_n} s_{11n}^{\prime 2}}} |d\bar{\varepsilon}_{11}^{pl}| = \frac{N_0 \, K_\tau}{v_e \left\{ \sqrt{c^2 - \frac{1}{\eta_n} s_{11n}^{\prime 2}} \right\}} |\bar{\varepsilon}_{11}^{pl}|_{crit.} \tag{184}$$

where $\left\{ \sqrt{c^2 - \frac{1}{\eta} s_{11n}^{\prime 2}} \right\}$ is the average value of the root with regard to the integration.

For $N_\Delta = 1$ we can write:

$$|\bar{\varepsilon}_{11}^{pl}|_{crit} = \frac{v_e K_c \left\{ \sqrt{c^2 - \frac{1}{\eta_n} s_{11n}'^2} \right\}}{N_0 K_{\tau'}} = K \qquad (185)$$

For $N_\Delta > 1$ the value of $\left\{ \sqrt{c^2 - \frac{1}{\eta_n} s_{11n}'^2} \right\}$ will be higher than for $N_\Delta = 1$, as the average value of s_{11n}' is lower. This follows from the fact that in every cycle the value of s_{11n}' increases in the first half of the cycle and in the second half it drops to zero and then goes to the opposite sign. If the number of the cycles increases with the resulting value of $(\sigma_e^r)_{crit}$ held fixed, the maximal value that $|s_{11n}'|$ reaches, decreases. Hence, the value of $\left\{ \sqrt{c^2 - \frac{1}{\eta_n} s_{11n}'^2} \right\}$ increases. Therefore, for $N_\Delta > 1$ it results:

$$|\bar{\varepsilon}_{11}^{pl}|_{crit.} > K \qquad (186)$$

similarly as in (183).

I. 1. 12 The Solution to the Identification Problem Based on the Flow-Curve

In the case of time-dependent processes the experimental input-data are usually represented by the flow-curves under constant load. The differential equation that describes the macroscopic flow-curve follows from eqs. (46) to (51). Whereas for the description of time-independent elastic-plastic processes it was suitable to use directly eqs. (46) to (51), here it seems preferable to exclude one of the two internal tensorial variables (σ_{ijn}') and introduce macroscopic strain $(\bar{\varepsilon}_{ij})$ instead. From eqs. (38) to (45) it can easily be derived:

$$s_{ijn}' = \frac{\eta_n}{\eta_e}\left(\frac{1+\eta_e}{v_e} \bar{s}_{ij} - \frac{v_n + \eta_e}{v_e} s_{ijn} - \frac{\bar{e}_{ij}}{\mu_e} \right) \qquad (187)$$

$$\sigma_n' = \frac{\eta_n^0}{\eta_e^0}\left(\frac{1+\eta_e^0}{v_e} \bar{\sigma} - \frac{v_n + \eta_e^0}{v_e} \sigma_n - \frac{\bar{\varepsilon} - \alpha_e T}{\varrho_e} \right) \qquad (188)$$

/This step can be done only due to the elasticity of the e-material. /

Using the above expressions we can rewrite eqs. (46) and (47) in the following form:

$$\dot{\bar{e}}_{ij} = \bar{\mu} \dot{\bar{s}}_{ij} + \frac{v_n}{R}\left\{ (\mu_n - \mu_e)\eta_n \left[\mu_e (1+\eta_e) \bar{s}_{ij} - v_e \bar{e}_{ij} \right] + \right. \qquad (189)$$

$$+\mu_e^2(\eta_e\dot{\eta}_n+v_e\eta_e+v_n\eta_n)s_n\Big\}\dot{h}$$

$$\dot{\bar{\varepsilon}}=\bar{\varrho}\dot{\bar{\sigma}}+\frac{v_n}{R_0}\Big\{(\varrho_n-\varrho_e)\eta_n^0\left[\varrho_e(1+\eta_e^0)\bar{\sigma}-v_e(\bar{\varepsilon}-\alpha_e T)\right]+$$

$$+\varrho_e^2(\eta_e^0\eta_n^0+v_e\eta_e^0+v_n\eta_n^0)\sigma_n\Big\}\dot{\varrho}_n+\bar{\alpha}\dot{T}+v_n M_0\dot{\tau}$$

$$\dot{s}_{ijn}=M\dot{\bar{s}}_{ij}+\frac{1}{R}\Big\{\eta_n\left[\mu_e(1+\eta_e)\bar{s}_{ij}-v_e\bar{e}_{ij}\right]- \tag{190}$$

$$-\left[\mu_e\eta_e\eta_n+v_e^2\mu_e\eta_e+v_n(\mu_e+v_e\mu_n)\eta_n\right]s_n\Big\}\dot{h}$$

$$\dot{\sigma}_n=M_0\dot{\bar{\sigma}}+\frac{1}{R_0}\Big\{\eta_n^0\left[\varrho_e(1+\eta_e^0)\bar{\sigma}-v_e(\bar{\varepsilon}-\alpha_e T)\right]-$$

$$-\left[\varrho_e\eta_e^0\eta_n^0+v_e^2\varrho_e\eta_e^0+v_n(\varrho_e+v_e\varrho_n)\eta_n^0\right]\sigma_n\Big\}\dot{\varrho}_n+$$

$$+N_0(\alpha_e-\alpha_n)\dot{T}-N_0\dot{\tau}$$

where $\bar{\mu}$, R, $\dot{h}$, $\bar{\varrho}$, R_0, M, M_0, N_0, $\bar{\alpha}$ are defined by eqs. (48) to (53).

We will limit ourselves to the case, in which the inelastic material constituent is described as a Maxwell body, i.e. as a material with immediate elastic response followed by linear viscous flow. This means that

$$\dot{h}=\frac{1}{2H}=const. \tag{191}$$

We use eq. (191) in $(189)_1$ and $(190)_1$, make the second derivative in eq. $(189)_1$ and exclude $\dot{s}_{ijn}$ and s_{ijn}. In this way the respective macroscopic rheological equation without internal variables is derived:

$$\ddot{\bar{e}}_{ij}=\bar{\mu}\,\ddot{\bar{s}}_{ij}+A\dot{\bar{s}}_{ij}+B\bar{s}_{ij}+C\dot{\bar{e}}_{ij}+D\bar{e}_{ij} \tag{192}$$

with

$$A=\frac{\mu_e}{2HR}\left[(v_e\mu_e+2v_n\mu_n)\eta_e\eta_n+v_e\mu_e\eta_e+\right. \tag{193}$$

$$+ 2v_n\mu_n\eta_n\Big] = 2H\mu_n B + \frac{\bar{\mu}}{2H\mu_n}$$

$$B = \frac{v_n\mu_e\eta_n}{4H^2R}(1+\eta_e)$$

$$C = -\frac{1}{2HR}\cdot\left[\mu_e\eta_e\eta_n + v_e^2\mu_e\eta_e + v_n(v_n\mu_e + 2v_e\mu_n)\eta_n\right]$$

$$D = -\frac{v_e v_n \eta_n}{4H^2R}$$

The solution to the identification problem is now easy. From the flow curve under constant load /macroscopic stress/ it is possible to determine the structural parameters η_e, η_n and the material constant H. The other constants $v_e = 1 - v_n$, μ_e , μ_n are assumed to be known /one of them can be determined from the easily measurable value of $\bar{\mu}$ using an iterative procedure/.

In the case of constant macroscopic stress the two first addends on the right-hand side of eq. (192) vanish / $\dot{\bar{s}}_{ij} = \ddot{\bar{s}}_{ij} = 0$ /. The value of $\bar{s}_{ij}$ is known. We measure the values of $\ddot{\bar{e}}_{ij}$, $\dot{\bar{e}}_{ij}$, $\bar{e}_{ij}$ at three selected points on the flow curve and using them in eq. (192) we get three linear equations for three unknowns B, C, D. With the values of B, C, D calculated we express the sought constants:

From

$$\frac{B}{D} = \mu_e \frac{1+\eta_e}{v_e} \tag{194}$$

it is possible to calculate η_e.

From

$$\frac{C}{D} = 2H\frac{\mu_e\eta_e\eta_n + v_e^2\mu_e\eta_e + v_n(v_n\mu_e + 2v_e\mu_n)\eta_n}{v_e v_n \eta_n} \tag{195}$$

we express H as a function of η_n and using this expression in $(193)_2$ we arrive at a quadratic equation for η_n, which solves the identification problem.

A disadvantage of this approach is the necessity of measuring the second derivative $\ddot{\bar{e}}_{ij}$ on an experimental curve, which is difficult to perform with a satisfactory precision. Therefore, if some special features of the problem under study can simplify the analysis, it is recommendable to look for another approach, as will be shown in a succeeding paragraph.

The problem of determining the structural parameters η_e^0, η_n^0

related to the isotropic parts of the stress- and strain- tensors was discussed at the end of paragraph I.1.10 and its solution is the same here.

I.1.13 Applications to Concrete

I.1.13.1 Creep of concrete

It is well known that the rheological properties of concrete can satisfactorily be described by the "Burgers model" that is characterized by the following macroscopic rheological equation:

$$\ddot{e}_{ij} = \bar{\mu}\,\ddot{\bar{s}}_{ij} + A\,\dot{\bar{s}}_{ij} + B\,\bar{s}_{ij} + C\,\dot{e}_{ij} \qquad (196)$$

This finding was arrived at in a phenomenological way, as a description of the observed macroscopic behaviour.

In our concept the same equation can be derived from an insight into the structure. The structure of concrete can approximately be modelled as elastic inclusions in a matrix that has immediate elastic response and subsequent creep, which - if described in the simplest way - can be assumed linear. Hence, the matrix has the properties of the "Maxwell body". Therefore, creep of concrete is expected to be describable by eqs. (189) to (193) with $\hbar = 1/2H$, $\eta_e = \eta_e^0 = \infty$, $\dot{\varrho}_n = T = \dot{T} = \dot{\tau} = 0$. This is really so, as the infinite value of η_e leads to $D = 0$ in eq. (192) whereas the other coefficients remain finite. /Similarly for elastic inclusions in a purely viscous matrix we have $\eta_e = \infty$, $\mu_n = 0$, which leads to vanishing values of the coefficients $\bar{\mu}$, D in eq. (192), i.e. to the rheological equation of the Lethersich type - generally used for the description of colloidal solutions./

Thus. eq. (192) changes its form to (196), where $\bar{\mu}$, A, B, C are defined by eqs. (68) and (193) where η_e is infinite. Similarly eqs. (189) and (190) must be accomodated for $\hbar = 1/2H$, $\eta_e = \eta_e^0 = \infty$, $\dot{\varrho}_n = T = \dot{T} = \dot{\tau} = 0$. Furthermore, it seems advantageous to use σ_{ije} instead of σ_{ijn} for the internal variable. This is easy to perform with the use of eq. (1):

$$\sigma_{ije} = \frac{1}{v_e}\left(\bar{\sigma}_{ij} - v_n\,\sigma_{ijn}\right)$$

and we arrive at the following set of equations:

$$\dot{e}_{ij} = \bar{\mu}\left(\dot{\bar{s}}_{ij} + \frac{1}{2H\mu_n}\,\bar{s}_{ij}\right) - v_e\,\mu_e\,\frac{v_e + \eta_n}{2HR}\,s_{ije} \qquad (197)$$

$$\dot{s}_{ije} = \mu_n \frac{v_e + \eta_n}{R} \dot{\bar{s}}_{ij} + \frac{v_e + \eta_n}{2HR} \bar{s}_{ij} - \frac{v_e^2 + \eta_n}{2HR} s_{ije}$$

$$\dot{\bar{\varepsilon}} = \bar{\varrho} \dot{\bar{\sigma}}$$

$$\dot{\sigma}_e = \varrho_n \frac{v_e + \eta_n^0}{R_0} \dot{\bar{\sigma}}$$

where $\bar{\mu}$, $\bar{\varrho}$, R , R_0 are defined by eqs. (68) to (72) .

To solve the problem of identifying the parameters of the mathematical model on the creep curve we concentrate our attention to the deviatoric parts at first. A sample is supposed to be subjected to an instantaneous compression $\bar{\sigma}_{11} = const.$. It follows from eqs. (197) that in this case all deviatoric quantities are similar to deviator $\bar{s}_{ij}$, i.e.

$$\bar{s}_{ij} = \bar{s}\, T_{ij} \;, \quad \bar{e}_{ij} = \bar{e}\, T_{ij} \;, \quad s_{ije} = s_e\, T_{ij} \qquad \text{etc.}$$

where

$$\bar{s} = \bar{s}_{11} = \frac{2}{3} \bar{\sigma}_{11}$$

and T_{ij} is constant, given in eq. (101) .

In the course of creep $\dot{\bar{s}}_{ij} = 0$ and eqs. $(197)_{1,2}$ can be rewritten in the form:

$$\begin{aligned} \dot{\bar{e}} &= A\bar{s} - Bs_e \\ \dot{s}_e &= C\bar{s} - Ds_e \end{aligned} \tag{198}$$

where

$$A = \frac{1}{2Hb} \;, \quad B = \frac{v_e \alpha (v_e + \eta_n)}{2H[\eta_n + v_e(v_e + v_n \alpha)]} \;, \tag{199}$$

$$C = \frac{B}{v_e \alpha \bar{\mu} b} \;, \quad D = C \frac{v_e^2 + \eta_n}{v_e + \eta_n}$$

$$b = \mu_n / \bar{\mu} \;, \qquad \alpha = \mu_e / \mu_n$$

According to (197) , at the moment of applying the load, $\bar{e}$ and s_e change over zero values to

$$\bar{e}^{0+} = \bar{\mu}\, \bar{s} \tag{200}$$

$$s_e^{0+} = \mu_n \frac{v_e + \eta_n}{\mu_n \eta_n + v_e (v_e \mu_n + v_n \mu_e)} \bar{s}$$

Integrating eq. $(198)_2$ we obtain

$$s_e = \frac{1}{D}\left[C\bar{s} - (C\bar{s} - Ds_e^{0+})\, e^{-Dt}\right] \tag{201}$$

Substituting (201) in $(198)_1$ and subsequent integrating results in:

$$\bar{e} - \bar{e}^{0+} = \bar{s}\left(A - \frac{BC}{D}\right)t - \frac{B}{D^2}\left(C\bar{s} - Ds_e^{0+}\right)\left(e^{-Dt} - 1\right) \tag{202}$$

The expression e^{-Dt} is eliminated from eq. (202) by the use of eqs. (201) and $(198)_1$; after rearranging we obtain

$$D(\bar{e}/\bar{s} - \bar{\mu}) = M + Nt - \dot{\bar{e}}/\bar{s} \tag{203}$$

where

$$M = A - B\frac{s_e^{0+}}{\bar{s}} = \frac{v_n}{2Hb} \cdot \frac{b\eta_n + v_e \alpha}{\eta_n + v_e(v_e + v_n \alpha)} \tag{204}$$

$$N = AD - BC = \frac{v_n \eta_n}{4H^2 \bar{\mu} b\left[\eta_n + v_e(v_e + v_n\alpha)\right]}$$

From eq. (203) a system of three linear equations can be formulated for determining the values of D , M , and N. The values of $\bar{\mu}$ and $\bar{s}$ are known. Furthermore, we select three points on the creep diagram and measure on them t , $\bar{e}$, and $\dot{\bar{e}}$. Inserting into eq. (203) gives the three linear equations. With these values of D, M, N known we can derive a quadratic equation for α /starting from eqs. (68) , (199) , (204) /:

$$\alpha^2 + \frac{\alpha}{v_e}\left[v_n\left(1 + \frac{1+K}{1 - \frac{MK}{\bar{\mu}D}}\right) + \frac{1+v_e}{K}\right] - \frac{1}{K}\left(1 + \frac{1+K}{1 - \frac{MK}{\bar{\mu}D}}\right) = 0 \tag{205}$$

where

$$K = \frac{D}{2(DM - N)}\left[-M + D\bar{\mu} \pm \sqrt{(M - D\bar{\mu})^2 + 4\bar{\mu}(DM - N)}\right] \tag{206}$$

$$\eta_n = v_e(v_n \alpha K - v_e) \tag{207}$$

$$2H = \frac{v_n \eta_n D}{N(v_e^2 + \eta_n)} \qquad (208)$$

$$b = \frac{v_n \eta_n}{4H^2 N\bar{\mu}\left[\eta_n + v_e(v_e + v_n \alpha)\right]} \qquad (209)$$

$$\mu_n = \bar{\mu} b \qquad (210)$$

$$\mu_e = \mu_n \alpha \qquad (211)$$

which solves the identification problem, supposing that v_e, v_n are known. This, however, is a rather complicated question in the case of concrete. There are grains of very different size present and their stress, strain and mechanical influence differs so much that they can hardly be described as one phase, as one material constituent. The reason for this difference lies in the overall heterogeneity of the stress field and strain field. One grain in a homogeneous and homogeneously stressed unlimited medium has some corresponding stress- and strain- field inside, which does not depend on the size of the grain. However, if the stress field in the surrounding medium is heterogeneous, then those grains, which are small with regard to the characteristic wavelength of the surrounding heterogeneous stress field. are stressed approximately in the same way as in a homogeneous surrounding stress field. On the other hand the grains that are large resist also the heterogeneous fluctuations of the field and therefore, their specific elastic energy content is higher.

Hence, it seems to be adequate to consider only the coarse fraction of the filler as the elastic inclusions and the remainder - the fine fractions - as a part of the viscoelastic matrix. Of course, it is not easy to decide, where is the boundary line.

We could select $v_e = 1 - v_n$ arbitrarily in the interval given by the complete volume of all filler fractions, but:

a/ The solution need not necessarily exist /and it was verified that for high values of v_e it does not exist in some cases/.

b/ If the solution exists, it describes all right the experimental flow-curve according to eq. (203) and the experimental value of $\bar{\mu}$, but the resulting values of μ_e, μ_n can be quite false.

It means that the input information is not sufficient, v_e cannot be arbitrary. The information that is stable and relatively easy to arrive at is the value of μ_e corresponding to the elastic inclusions. We proceed in such a way that different values of v_e are subsequently tried in eqs. (205) to (211) until such value of ν_e is found. for which the resulting value of μ_e /eq. (211) / equals the known value of μ_e. With a small computer this is easy to perform.

J. Jírová and V. Kafka [24] used a bit different procedure, but that explained here above seems now preferable.

For a numerical example the experimental data published by F. Kruml [43] were used. The material was characterized by the following original data:

$$\bar{E} = 4 \cdot 10^4 \; MPa \qquad \bar{\mu} = 2.88 \cdot 10^{-5} \; (MPa)^{-1}$$

$$\bar{\nu} = 0.15 \qquad \bar{\varrho} = 1.75 \cdot 10^{-5} \; (MPa)^{-1}$$

The mean values of the constants D, M, and N were obtained from two triplets of arbitrarily chosen points on the creep curve. The measured values of $t, \bar{e}, \dot{\bar{e}}$ at these points were used in eq. (203), which resulted in:

$$D = 9.409 \cdot 10^{-2} \; (day)^{-1} \qquad N = 1.715 \cdot 10^{-8} (MPa)^{-1} (day)^{-1}$$

$$M = 3.965 \cdot 10^{-6} \; (MPa)^{-1} (day)^{-1}$$

The properties of the elastic inclusions were considered:

$$E_e = 49 \; GPa$$

$$\nu_e = 0.079$$

which means

$$\mu_e = (1 + \nu_e)/E_e = 2.2 \cdot 10^{-5} \; (MPa)^{-1}$$

The value of ν_e that corresponds to it in eqs. (205) to (211) is $\nu_e = 0.309$.
and the other resulting material constants are:

$$\mu_n = 3.32 \cdot 10^{-5} \; (MPa)^{-1}$$

$$H = 6.87 \cdot 10^4 \; (MPa)(day)$$

$$\eta_n = 3.583 \cdot 10^{-3}$$

For the determination of σ_e, σ_n we assume that the volume changes are merely elastic, which means that σ_e, σ_n will not change with time. The structural parameter η_n^0 is approximately calculated from eqs. (128) and (70) that are looked upon as two equations for two unknowns ϱ_n, η_n^0. Solution gives:

$$\eta_n^0 = 0.2908$$

which means that

$$\varrho_n = 1.76 \cdot 10^{-5} (MPa)^{-1}$$

$$\nu_n = 0.185$$

With these parameters determined and with the use of eqs. (200), (201), (202), (1), $(42)_{1,2}$, $(197)_{3,4}$, (2), it is easy to plot the stress- and strain- diagrams in Fig. 18.

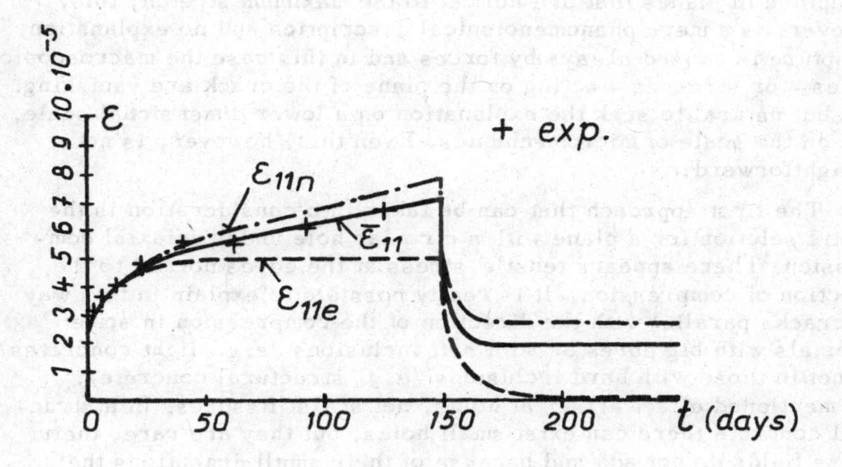

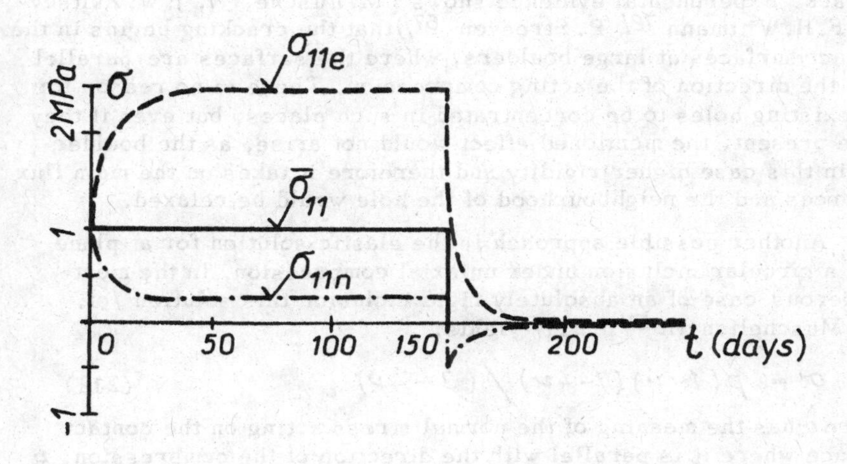

Fig. 18 Creep of concrete - internal stresses and strains

I.1.13.2 Cracking under compression

Compressive loading of concrete or similar materials leads to the creation of cracks that are parallel with the direction of the acting compression, or - if the compression acts in two directions - with these two directions. The understanding and mathematical modelling of this phenomenon is not trivial. It is possible to formulate the criterion of rupture in planes that are normal to the maximum stretch, this, however, is a mere phenomenological description and no explanation. A rupture is caused always by forces and in this case the macroscopic forces - or stresses - acting on the plane of the crack are vanishing. It is but natural to seek the explanation on a lower dimensional scale, i.e. on the scale of micromechanics. Even this, however, is not straightforward:

The first approach that can be taken into consideration is the elastic solution for a plane with a circular hole under uniaxial compression. There appears tensile stress at the edges normal to the direction of compression. It is really possible to explain in this way the cracks parallel with the direction of the compression in some materials with big pores or with soft inclusions /e.g. light concretes/, but not in those with hard inclusions /e.g. structural concrete/. The mentioned effect arises at holes, not at flat fissures. In a structural concrete there can exist small holes, but they are rare, their stress fields do not add and because of their small dimentions the energy concentrated at one hole is not sufficient for the creation of a crack. Experimental evidence shows (M. Lusche [47/], J.W. Zaitsev and F.H. Wittmann [78/], P. Stroeven [67/])that the cracking begins in the contact surfaces at large boulders, where the surfaces are parallel with the direction of the acting compression. There is no reason for the existing holes to be concentrated in such places, but even if they were present, the mentioned effect would not arise, as the boulder has in this case higher rigidity and therefore it takes on the main flux of forces and the neighbourhood of the hole would be relaxed.

Another possible approach is the elastic solution for a plane with a circular inclusion under uniaxial compression. In the most dangerous case of an absolutely rigid inclusion this solution /cf. N.J. Muschelishvili [51/]p.211/ yields:

$$\sigma = p(1-\nu)(1-4\nu) \Big/ (3-4\nu) \qquad (212)$$

where σ has the meaning of the normal stress acting on the contact surface where it is parallel with the direction of the compression, p is the compressive stress loading the plane and ν is the Poisson ratio of the plane. For $\nu = 0.2$, which corresponds to the cement stone, the value of σ turns out to be 0.073p, i.e. a very low compressive stress. The analogous solution in three dimensions does not lead to significant values either /cf. M. Lusche [47]/. For inclusions that differ in their shape from a circle the stresses can be a bit higher /cf. R.H. Edvards [11]/, but the difference is not substantial and experimental evidence does not show /cf. J.W. Zaitsev and F.W. Wittmann [78], T.C. Thomas [70], S.P. Shah and F.O. Slate [64]/ that the cracks would be concentrated at elongated shapes of grains. It does not seem therefore that the microstresses arising in the elastic range could be the cause of the cracks under consideration.

In Stroeven's monograph [67] some papers are quoted, where the cause of the cracks parallel with the direction of the acting compression is seen in the interaction of spheres /particles of the aggregates/. This explanation does not seem to be satisfactory either. It is possible to imagine two spheres and another one that is pressed between them and in this way they are separated. But it is also possible to imagine another configuration by which they are brought nearer. If we imagine a regular arrangemenr of spheres that are smooth and lubricated and boundaries that are parallel with the direction of the acting compression and free of stress, then the spheres in the planes normal to the direction of the compression will be separated. However, if there exists cohesion among the spheres, this effect will be a boundary effect, maximum at the free boundaries and diminishing with the distance from them. According to this scheme the cracking should appear at free boundaries by breaking out the boundary boulders. This, however, is not observed as a rule. Another argument against this explanation is the fact, that observations of slices of cracked concrete do not reveal concentration of cracking in places, where one boulder is pressed between two others.

The key for the understanding of the real cause of cracking in the planes parallel with the acting compression is after all eq. (212). For Poisson's ratio 0.5 the value of the stress resulting from it is -0.5p, i.e. a relatively high tensile stress. Poisson's ratio of the cement stone is 0.2 and therefore this result is of no meaning for the elastic range, but the value 0.5 is characteristic for a process, in which the volumetric changes are zero and the deformation has only deviatoric character. This is the characteristic feature of a plastic or viscous process. In the case of a quasihomogeneous microfracturing of the matrix volumetric changes appear, but they are of an opposite sign - dilatancy - , which leads to an opposite effect - rise of the tensile stress.

On the basis of the considerations outlined above we are going to formulate the basic hypotheses of our approach:

a/ In the case of an uniaxial compressive loading the cracks originate at the contact surfaces of large boulders of the hard aggregates in places, where these surfaces are parallel with the direction of the acting compression.

b/ The cause of these cracks are tensile stresses arising as a consequence of an inelastic deformation, which can be plastic, viscous, or can have the character of quasihomogeneous stable microfracturing.

c/ In the case of a compressive loading in two directions the process is quite analogous, the cracks are parallel with the plane given by the two directions of the acting compressions.

These basic hypotheses are strongly corroborated by Stroeven's experimental results. According to his findings /cf. 67/ / uniaxial compression leads to the separation of biconical forms with a boulder of the aggregates in their centre - cf. Fig. 19.

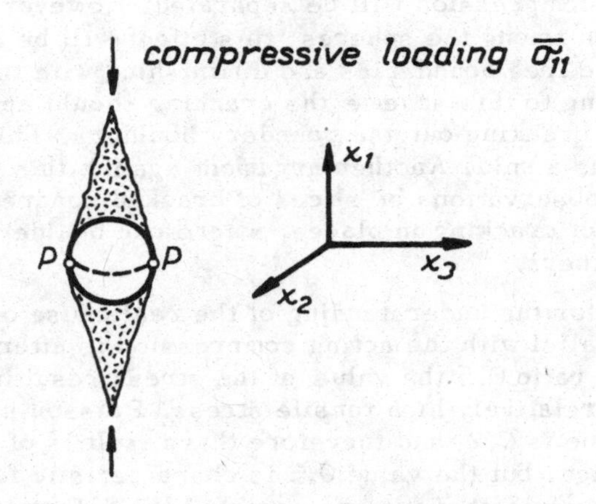

Fig. 19 Model of a grain with two conical formations separated from the matrix by the fracturing process under uniaxial compression.

The cracking begins at the surface of the boulder /this evidence is supported also by a number of other experimental works - e.g. 47/, 78/, 70/, 64/ / in places, where the surface is parallel with the direction of the acting compression. It is natural to model these forms as geometric forms with rotational symmetry and then the stress causing the beginning of cracking can be only tensile stress, as due to the symmetry the shear stresses are zero. It was shown in the preceding considerations that these tensile stresses can have significant values only as a consequence of inelastic deformation.

In this context we will model structural concrete - or any similar material - as a two-phase composite of elastic inclusions in a matrix that displays elastic deformation and some kind of inelastic deformation - plastic, viscous, or quasihomogeneous stable microfracturing. The adequate mathematical model is represented by eqs. (66) to (72). It follows from these equations that for a space - independent macroscopic stress in the course of the whole loading history the stress-field in the inclusions is also space - independent /which is not the case in the matrix/. The knowledge of the stress field in the inclusion is sufficient for the determination of the contact stresses t_i :

$$t_i = \sigma_{ije} \, n_j \tag{213}$$

where n_j is the outward unit normal of the surface of the inclusion and σ_{ije} is the tensor of the homogeneous stress field in the inclusion.

In accordance with Fig.19 the inclusion is modelled as a sphere and the contact stress at all the equatorial points P will be equal. Let us choose point $P^{(2)}$, where the normal is parallel with the coordinate axis x_2. The contact stress in this point will be:

$$t_i^{(2)} = \sigma_{ije} \, n_j^{(2)} = \sigma_{i2e} \tag{214}$$

as the only non-vanishing component of $n_j^{(2)}$ is $n_2^{(2)} = 1$.

The loading under study is a compressive uniaxial macrostress in the x_1-direction and in this case it is advantageous to describe the deviatoric part as follows:

$$\bar{s}_{ij} = \bar{s} \, T_{ij} \tag{215}$$

where $\bar{s} = \bar{s}_{11}$ and the tensor T_{ij} is given in eq. (101).

It is a consequence of eqs. (66), (67) that in this case all the deviatoric components of stresses and strains with unequal indices $(i \neq j)$ are zero and those with equal indices can be expressed as products of some scalars with the tensor T_{ij}, e.g. $s_{ije} = s_e T_{ij}$ etc.

Hence, the only non-vanishing component in eq. (214) will be

$$t_2^{(2)} = \sigma_{22e} = s_{22e} + \sigma_e \tag{216}$$

and due to the form of the tensor T_{ij} it holds:

$$s_{22e} = -\frac{1}{2} s_{11e} = -\frac{1}{2} s_e \tag{217}$$

where s_e can be expressed with the use of eq. (1) in the following way:

$$s_e = \frac{1}{v_e} (\bar{s} - v_n s_n) \tag{218}$$

In equations (66), (67) we can clearly distinguish the elastic response and the inelastic parts. It can be shown /cf. Appendix (5)/ that for a uniaxial monotonic macroscopic loading the sign of the inelastic part of $\dot{s}_e$ is the same as that of the elastic part of $\dot{s}_e$ and that of $\dot{\bar{s}}$. This means:

$$\begin{gathered} \dot{s}_e = (\dot{s}_e)_{el} + (\dot{s}_e)_{nl} \\ sign\,(\dot{s}_e)_{nl} = sign\,(\dot{s}_e)_{el} = sign(\dot{s}_e) = sign(\dot{\bar{s}}) \end{gathered} \tag{219}$$

The isotropic components will be given only by the elastic response in the case of plastic or viscous deformation. In the case of quasi-homogeneous microfracturing there are two addends in the inelastic part of $\dot{\sigma}_e$ /given by eq. $(67)_2$ with $\dot{T}=0$ and an equation that is quite analogous to eq. (218)/. For the first addend, connected with $\dot{\varrho}_n$ the signs will be the same as in the case of $\dot{s}_e$. But the second addend, connected with $\dot{\tau}$ usually overweights, which manifests itself by the fact that the volume increases in spite of the compressive loading. The sign of the second addend is always positive /cf. Appendix 5/ and therefore, we have:

$$\begin{gathered} \dot{\sigma}_e = (\dot{\sigma}_e)_{el} + (\dot{\sigma}_e)_{nl} \\ (\dot{\sigma}_e)_{nl} \geq 0 \end{gathered} \tag{220}$$

For a uniaxial compressive monotonic macroscopic loading eq. (216) can now be rewritten as follows:

$$\begin{aligned} t_2^{(2)} &= (s_{22e})_{el} + (s_{22e})_{nl} + (\sigma_e)_{el} + (\sigma_e)_{nl} = \\ &= (t_2^{(2)})_{el} - \frac{1}{2}(s_e)_{nl} + (\sigma_e)_{nl} \end{aligned} \tag{221}$$

with

$$-\frac{1}{2}(s_e)_{nl} > 0 \;, \quad (\sigma_e)_{nl} \geq 0$$

which means that the contact stresses at the equatorial points P are changed by the inelastic deformation and this change has the sign of tension.

The numerical example that follows shows that the influence of the inelastic deformation upon the contact stresses is significant.

This numerical example describes the changes of internal stresses due to creep as it is partly presented in the preceding paragraph I. 1. 13. 1. Whereas in the preceding paragraph the analysis was aimed at the course of the average stresses in the material constituents, here we are interested in the contact stresses according to eq. (216).

For the deviatoric elastic response of s_{11e} we get from eqs. (67) and (218):

$$(s_{11e})_{el} = (s_e)_{el} = \mu_n \frac{v_e + \eta_n}{\mu_n \eta_n + v_e(v_e \mu_n + v_n \mu_e)} \bar{s}_{11} \qquad (222)$$

Furthermore, it can be shown /cf. Appendix 5/ that after a long time /theoretically infinite time/ the value of s_{11e} will change to:

$$(s_{11e})_\infty = (s_e)_\infty = \frac{v_e + \eta_n}{v_e^2 + \eta_n} \bar{s}_{11} \qquad (223)$$

The response of the isotropic part σ_e will be only elastic. From eq. $(67)_2$ and eq. (1) /rewritten for isotropic parts/ we easily arrive at:

$$\sigma_e = \frac{\varrho_n (v_e + \eta_n^o)}{\varrho_n \eta_n^o + v_e(v_e \varrho_n + v_n \varrho_e)} \bar{\sigma} \qquad (224)$$

For the uniaxial compression $\bar{\sigma}_{11}$ it holds:

$$\bar{s}_{11} = \frac{2}{3} \bar{\sigma}_{11} \quad , \qquad \bar{\sigma} = \frac{1}{3} \bar{\sigma}_{11}$$

and inserting the numerical values from I. 1. 13. 1 we get:

$$(\sigma_e)_{el} = (\sigma_e)_\infty = 1.00738\, \bar{\sigma} = 0.3358\, \bar{\sigma}_{11}$$

$$(s_{11e})_{el} = (s_e)_{el} = 1.2958\, \bar{s}_{11} = 0.8639\, \bar{\sigma}_{11}$$

$$(s_{11e})_\infty = (s_e)_\infty = 3.1580\, \bar{s}_{11} = 2.1053\, \bar{\sigma}_{11}$$

and for the contact stresses at the equatorial points P according to eqs. (216) and (217):

$$\left(t_2^{(2)}\right)_{el} = -0.0962\, \bar{\sigma}_{11}$$

$$\left(t_2^{(2)}\right)_\infty = -0.7168\, \bar{\sigma}_{11}$$

Hence, we can see that in the elastic range the contact stresses are unimportant, but the inelastic deformation leads to considerable values of tensile stress.

The presented micromechanical model has meaning for such cases, where the redistribution of internal stresses due to inelastic deformation is possible, i.e. where it is not preceded by brittle fracture. For concrete and similar materials this is the case only in compression. However, for metals with resistant inclusions /e.g. precipitates/ the same model and the same equations can be used even for tension, where the cracks will be perpendicular to the direction of the acting loading. There exists experimental evidence that supports this scheme.

In the advanced model there does not appear any parameter characterizing the dimensions of the grains. The fact that the cracks appear preferably at large grains is explained by the "quantum theory of strength" in the last part of this book.

The process of an overall quasihomogeneous stable fracturing and the phenomenon of strain-softening is modelled in the following section.

I. 1. 13. 3 Quasihomogeneous stable microfracturing

The simplest yield conditions - the Mises criterion - is formulated in eq. (103). It can be understood as a hypersphere in the deviatoric stress space. i.e. the ability of the material to respond elastically is exhausted in a certain distance from the zero-stress--state.

An alternative interpretation of this criterion was given by H.Hencky /15/ . Equation (103) was rewritten as

$$s_{ij}\, e_{ij}^{el} = \mu s_{ij}\, s_{ij} = \frac{e_{ij}^{el}\, e_{ij}^{el}}{\mu} \leq 2K^2 \qquad (225)$$

with $\mu = const.$

Equation(225) means that the deviatoric elastic energy that can be comprised in a unit volume of some material is limited.

Such a statement can be expressed in other words: the material has a limited elastic capacity, i.e. its elastic energy *(EE)* cannot overpass its elastic capacity *(EC)*:

$$EE \leq EC \qquad (226)$$

For many cases the deviatoric part of the elastic energy $(EE^{(d)})$ is decisive. i.e.

$$EE^{(d)} \leq EC^{(d)} \qquad (227)$$

for other cases also the tensile isotropic elastic capacity decides:

$$EE^{(i)} \leq EC^{(i)} \qquad (228)$$

This concept saying that elastic capacity has certain limits, similarly as other capacities in physics. seems to be of deep meaning, not only for plasticity.

After the elastic energy reached the elastic capacity two different processes can set in:
either

$$d(EE) = 0$$

or

$$d(EE) < 0$$

The first kind will be called active process, the second kind passive process or unloading.

Now let us discuss the ways in which the material can behave in an active process. We will call such processes post-critical. We are going to specify the fundamental simple mathematical models - the behaviour of a real material can often be described as a combination of them. For simplicity of the explanation we will limit ourselves at first to cases where only the deviatoric energy decides. Then according to equation (227) it holds in the post-critical state:

$$0 = \frac{d}{dt} EE^{(d)} = \frac{d}{dt}(s_{ij}^E e_{ij}^E) = \frac{d}{dt}\left[(s_{ij} - s_{ij}^L)(e_{ij} - e_{ij}^P)\right] = \tag{229}$$

$$= (\dot{s}_{ij} - \dot{s}_{ij}^L)(e_{ij} - e_{ij}^P) + (s_{ij} - s_{ij}^L)(\dot{e}_{ij} - \dot{e}_{ij}^P) =$$

$$= \dot{s}_{ij}^E \mu s_{ij}^E + s_{ij}^E (\dot{\mu} s_{ij}^E + \mu \dot{s}_{ij}^E) =$$

$$= (\dot{e}_{ij}^E/\mu - \dot{\mu} e_{ij}^E/\mu^2) e_{ij}^E + e_{ij}^E \dot{e}_{ij}^E/\mu$$

where the superscript E means "elastic", L means "locking" and P means "plastic":

$$e_{ij}^E = \mu s_{ij}^E$$

$$s_{ij}^L = s_{ij} - s_{ij}^E$$

$$e_{ij}^P = e_{ij} - e_{ij}^E$$

The best known and elaborated model with limited EC is the perfectly plastic model /PPM/, for which the Mises criterion was originally formulated.

In our concept the model is characterized by

$$\boxed{\dot{s}_{ij} = \dot{s}_{ij}^L = 0} \tag{230}$$

From (230) and (229) we get:

$$\dot{s}_{ij}^E = \dot{s}_{ij}^L = \dot{\mu} = 0\,, \quad s_{ij} = s_{ij}^E\,, \quad \dot{e}_{ij} = \dot{e}_{ij}^P \tag{231}$$

or in words:

In the post-critical process there is no change of stress, no change of the elastic constant, and the change of strain is only plastic. The respective stress-strain diagram is given in Fig. 20

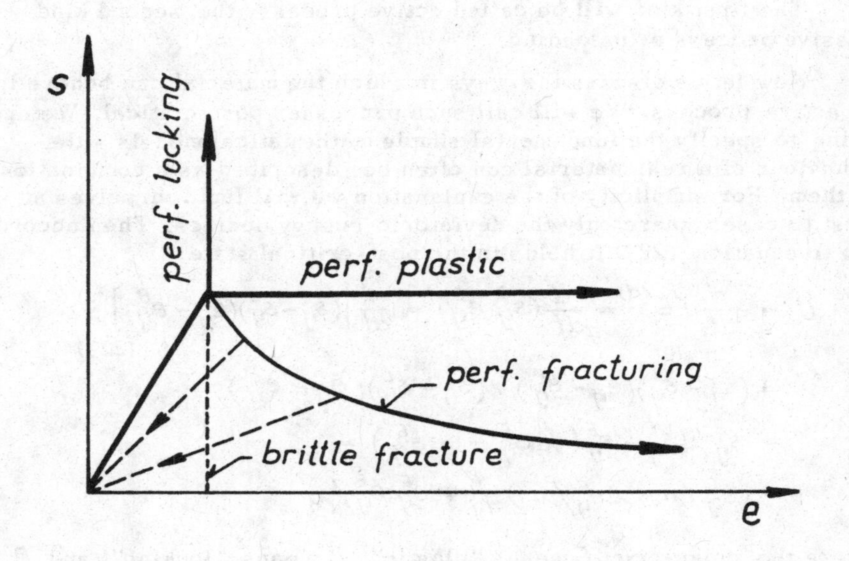

Fig. 20. Stress-strain diagrams of basic models of materials with limited elastic capacity.

A model that is dual to the above mentioned is called here a perfectly locking model /PLM/ /it is the simplest special case of what is described as "a perfectly locking model" in W. Parger 60/.

$$\boxed{\dot{e}_{ij} = \dot{e}^{P}_{ij} = 0} \tag{232}$$

From (232) and (229) it follows:

$$\dot{e}^{E}_{ij} - e^{P}_{ij} - \dot{\mu} - 0 \ , \quad e_{ij} = e^{E}_{ij} \ , \quad \dot{s}_{ij} = \dot{s}^{L}_{ij} \tag{233}$$

or in words:

In the post-critical process there is no change of strain, no change of the elastic constant, and the change of stress is only locking.

The respective stress-strain diagram is given again in Fig. 20.

The above cases are well known and were only shortly characterized and included in the concept.

The formulation of the perfectly fracturing model /PFM/ is new. At first we are going to define the model and comments will follow. In our concept the respective definition reads:

$$\boxed{\dot{s}_{ij}^{L} = \dot{e}_{ij}^{P} = 0} \qquad (234)$$

From (234) and (229) it results:

$$\dot{s}_{ij}^{E} = \dot{s}_{ij} , \quad \dot{e}_{ij}^{E} = \dot{e}_{ij} , \quad s_{ij}^{E} = s_{ij} , \quad e_{ij}^{E} = e_{ij} \qquad (235)$$

or in words:

In the post-critical process the relation between stress and strain remains elastic, but the elastic constant changes due to stable homogeneous fracturing.

From (229), (234) and (235) we get:

$$\dot{\mu} = -2\mu \frac{s_{ij}\,\dot{s}_{ij}}{s_{kl}\,s_{kl}} = 2\mu \frac{e_{ij}\,\dot{e}_{ij}}{e_{kl}\,e_{kl}} \qquad (236)$$

For a monotonically increasing strain

$$e_{ij} = e\,T_{ij}$$

with T_{ij} constant it holds

$$s_{ij} = s\,T_{ij}$$

and equation (236) can be integrated:

$$\frac{s}{s_0} = \frac{e_0}{e} \qquad (237)$$

$$\frac{\mu}{\mu_0} = \frac{s_0^2}{s^2} = \frac{e^2}{e_0^2} \qquad (238)$$

The respective stress-strain diagram corresponding to (237) - - a hyperbola - is depicted in Fig. 20 with the respective unloading paths that correspond to (238).

There can be no question of proving the validity of the PFM, PPM or PLM. They are the simplest mathematical schemes with only one constant describing them - the EC - and may or may not be suitable for the description of the response of a real material - usually in combinations.

The PFM is the simplest scheme descriptive of the cumulative damage process. Its applicability is discussed in what follows.

At first let us answer the question what kind of material could be described by this last model. Such a material should be absolutely brittle and homogeneous - not only in the initial state, but also during the fracturing process. An infinite number of infinitely small microcracks should be homogeneously distributed in the material and no microcrack should grow quicker than the others. It is very clear that such a process cannot exist in reality, as it is highly unstable. Any small imperfection of any kind leads to concentration of the damage, to a catastrophic process - the brittle fracture /see Fig. 20/. Then the general concept of quasihomogeneity ceases to be valid, and our approach is not more adequate. Therefore, the PFM scheme can be of interest in such cases only where the stability of the material is raised. This can be the case if plastic or viscous deformation proceedes simultaneously with the fracturing process /stress concentrations are relaxed by plastic or viscous deformation/, or - in a heterogeneous material - by the presence of another non-fracturing material constituent /the fracturing constituent stresses relax during the process, and they are taken over by the non-fracturing constituent, which means stabilization/. In such cases the PFM scheme seems to be of high significance. An example, where both above mentioned stabilizing effects appear, is concrete and similar materials.

The meaning of the "stable progressively fracturing solids" /change of elastic constants and unloading pointing to the zero-point/ for the phenomenological description of some materials was recognized earlier /cf. e.g. J.W.Dougill 9/, Z.P.Bažant and S.S.Kim 2//.

The advantage of the presented way is that it starts from the description of the real structure of the material, that there is the possibility to describe the stresses and strains in the material constituents and that the material in question is defined by a set of scalar constants only.

Furthermore, it is possible to specify the degree of damage in the fracturing component, and supposing that we know the characteristics of the distribution of imperfections, it is possible to evaluate the state where the catastrophic process - the localization of damage and rupture - takes place:

Thus there opens the possibility to combine the "cumulative damage mechanics" - and the "fracture mechanics" - approach, similarly as it is done in J.Janson and J.Hult 22/.

In his classical work M.Reiner 63/ considers three fundamental material properties:

a/ elasticity. b/ viscosity, c/ plasticity.

It seems, however, that a more complete set represented by the simple models could read:

A. Model with unlimited EC:
 1/ PEM /perfectly elastic material/,

B. Model with vanishing EC:
 2/ PVM /perfectly viscous material/,

C. Models with limited EC:
 3/ PPM /perfectly plastic material/,
 4/ PLM /perfectly locking material/,
 5/ PFM /perfectly fracturing material/.

In what preceded the possibility of combining simple models to describe the behaviour of a real material was repeatedly mentioned. Such an approach is old and widely used /cf. e.g. 63/ /, and represents only one application of the very general and very successful idea of describing complex elements as combinations of simple basic elements /vector spaces, the FEM method etc. /. The disadvantage and limitation of its classical form was that only a parallel or series arrangement /homogeneous stress or homogeneous strain assumption/ was possible to use, which did not correspond to the real co-effect of microparticles. New possibilities open with our concept, where the infrastructure of any material constituent can be described as discontinuous /inclusions/ or continuous /matrix/ with special structural parameters that characterize the influence of the geometry of composition.

It can be looked upon as a special case of the internal variables approach, the microstresses in the material constituents having the meaning of internal variables, and in the case of the PFM scheme also the changing elastic moduli are other internal variables describing cumulative damage.

To clarify the meaning of the above considerations, let us construct a simple model of a heterogeneous material with the following properties:

a/ "The composite consists of a matrix of one material kind and inclusions of another material kind",

b/ "Inclusions are elastic"

c/ "Matrix is a perfectly fracturing material with additional permanent deformations":

The change of the deviatoric compliance μ_n represents here the measure of fracturing and damage. Both the deviatoric and the isotropic parts of the increment of permanent strain are qualified by the fracturing process, and are proportional to $\dot{\mu}_n$. The relative change of the isotropic compliance ϱ_n equals the relative change of the deviatoric compliance μ_n, as both have the same cause - the fracturing. It is easy to prove that this last assumption is equivalent to the assumption of non-changing elastic Poisson's ratio.

Hence, the adequate form of eq. (53) will be:

$$\dot{h} = \dot{\lambda} + \dot{\mu}_n = (1+\varkappa)\dot{\mu}_n \tag{239}$$

and for $\dot{\varrho}_n$ and $\dot{\tau}$ the following relations are assumed to hold:

$$\frac{\dot{\varrho}_n}{\varrho_n} = \frac{\dot{\mu}_n}{\mu_n} \tag{240}$$

$$\dot{\tau} = \omega\dot{\mu}_n \tag{241}$$

where $\varkappa$ and ω are material constants.

With expressions (239) to (241) used in equations (66) and (67) we have the corresponding mathematical model.

To complete the solution, it is necessary to express $\dot{\mu}_n$ in terms of the macroscopic stress increments. This can be done on the basis of our elastic capacity concept which postulates that $EE^{(d)}$ cannot overpass a certain limit and is constant in any active post-critical process.

In a small macroscopic neighbourhood of a point, $EE^{(d)}$ comprised in a unit volume of one material constituent is not expressed by formula (225) only, as there are fluctuations of microstresses. The respective formula - according to eq. (18) for the n-material is:

$$2EE_n^{(d)} = \mu_n\left(s_{ijn}s_{ijn} + \frac{1}{\eta_n}s'_{ijn}s'_{ijn}\right) \leq 2K^2 \tag{242}$$

and with the use of eq. $(65)_1$ this formula can be transformed to:

$$\frac{\mu_n}{v_e^2}\left[(v_e^2+\eta_n)s_{ijn}s_{ijn} + \eta_n\bar{s}_{ij}\bar{s}_{ij} - 2\eta_n s_{ijn}\bar{s}_{ij}\right] \leq 2K^2 \tag{243}$$

This expression is to be differentiated in regard to $\mu_n, s_{ijn}, \bar{s}_{ij}$. The resulting differential must be nil in a post-critical active process, and using $(67)_1$ and (239) we arrive at the sought formula for $\dot{\mu}_n$:

$$\dot{\mu}_n = \frac{-2v_e^2\mu_n(Ps_{ijn} + \bar{P}\bar{s}_{ij})\dot{\bar{s}}_{ij}}{[1-2\mu_n Q(1+\varkappa)]s_{kln}[(v_e^2+\eta_n)s_{kln} - 2\eta_n\bar{s}_{kl}] + [1-2\mu_n\bar{Q}(1+\varkappa)]\eta_n\bar{s}_{pq}\bar{s}_{pq}} \tag{244}$$

where summing is to be performed on the indices ij, kl, pq and P, $\bar{P}$, Q, $\bar{Q}$ are given by eqs. (69).

This completes the macroscopic constitutive equation. The properties of the material in question are defined by 10 scalar constants:

$$v_e (= 1 - v_n), \mu_e, \rho_e, \mu_n^*, \rho_n^*, \eta_n, \eta_n^o, \varkappa, \omega, K$$

where μ_n^*, ρ_n^* are the respective values of μ_n, ρ_n in the virgin state before the beginning of the fracturing process.

If the 10 constants are known, we can calculate the response for a given loading path: the course of macroscopic strain, the changes of internal stresses and strains, the changes of the moduli of the fracturing material and the degree of its damage.

For a numerical example a program for the HP - 20 calculator was prepared on the basis of the preceding equations. It was supposed that the heterogeneous material was compressed in the x_1 -direction.

The set of the 10 constants defining the material was chosen as follows:

$$
\begin{aligned}
v_e &= 0.4 \\
\mu_e &= 2.225 \times 10^{-5} (MPa)^{-1} \\
\rho_e &= 1.347 \times 10^{-5} (MPa)^{-1} \\
\mu_n^* &= 6.675 \times 10^{-5} (MPa)^{-1} \\
\rho_n^* &= 4.041 \times 10^{-5} (MPa)^{-1} \\
\eta_n &= 0.02 \\
\eta_n^o &= 0.02 \\
\varkappa &= 2.85 \\
\omega &= 5 \\
2K^2 &= 1.478 \times 10^{-2} MPa
\end{aligned}
$$

The resulting stress-strain diagram was depicted by the plotter of the calculator /Fig. 21/ with the unloading paths in several points and the volumetric deformation.

We can see a very good qualitative agreement with the time--independent behaviour of concrete in the stress-strain diagram, in the unloading paths and in the volumetric changes.

It was possible to find such material parameters that the stress--strain curve was very close to experimental points taken from D. C. Spooner and J. W. Dougill [65], which made the coincidence even more clear.

A generalization of the model to the description of time-dependent behaviour would be easy.

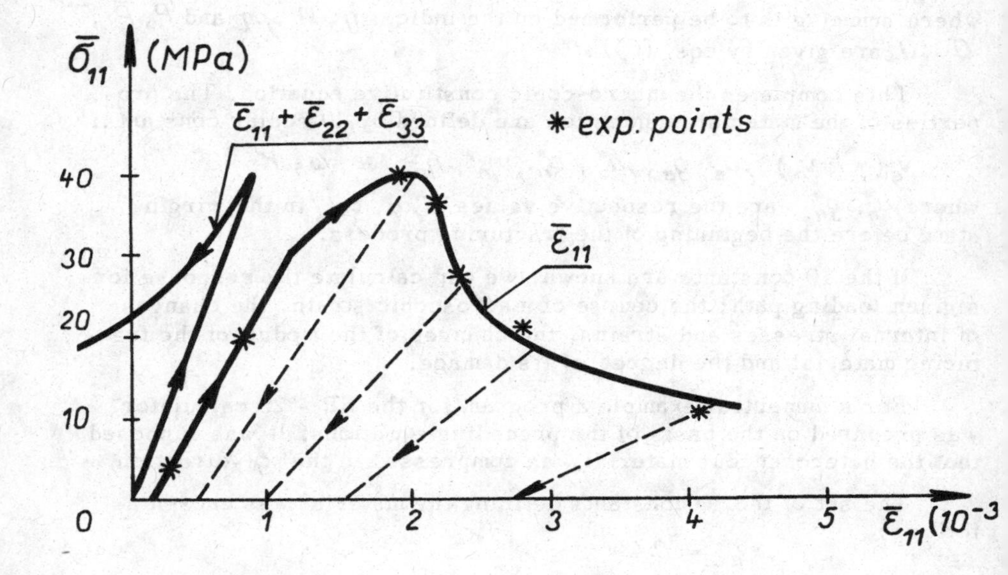

Fig. 21 Theoretical stress-strain diagram of a heterogeneous material with elastic inclusions in fracturing-plastic--dilating matrix; experimental points /taken from D. C. Spooner and J. W. Dougill [65]/ correspond to concrete.

It could seem surprising that the controlling formula for the elastic range and further process is in deviatoric components only /eq. (242)/. It seems, however, that for the range of compressive or vanishing first invariant of macroscopic stress tensor this is a reasonable approximation. For the case of tensile stresses the above simple scheme is not applicable not only because of the deviatoric criterion /which could be adapted/, but in the case of tensile loading it is not possible to describe the fracturing process by a scalar parameter only /here μ_n/, as it is strongly anisotropic in such cases /cf. [2]/. and so unstable that it is hardly possible to speak about quasihomogeneous fracturing at all.

I. 1. 14 Final Remarks on the Model for Materials with Macroscopic Isotropy

The structure of the model described in section I. 1 has some specific features that deserve attention.

First of all it is the separate response of the deviatoric and the isotropic parts on the mesoscale to the respective parts on the macroscale in the range of elastic deformation. It is clear enough that macroscopic deviatoric stress can cause also stresses with isotropic parts on the microscale and macroscopic isotropic stress can cause also microstresses with deviatoric parts. However, on the mesoscale the situation is different. To be concrete let us have in mind a two-phase material in elastic deformation with the only macroscopic stress component $\bar{s}_{12}$. According to our model the response on the mesoscale will be s_{12a}, s_{12b} with

$$\sigma_a = \sigma_b = 0$$

and $$s_{ija} = s_{ijb} = 0 \qquad \text{for} \qquad ij \neq 12$$

Firstly we will discuss the vanishing values of σ_a, σ_b. We are going to prove that they cannot be non-vanishing. To this purpose let us turn the coordinate axes X_1, X_2 by an angle of 45°, the axis of rotation being X_3. In the rotated system X_1^*, X_2^*, X_3 the only non-vanishing macrostress components will be: $\bar{\sigma}_{11}^* = \bar{s}_{12}$, $\bar{\sigma}_{22}^* = -\bar{s}_{12}$. For a moment let us assume that $\bar{\sigma}_{11}^*$ is acting separately and that the respective mesoscale isotropic response is $(\sigma_a)_1$, $(\sigma_b)_1$. Due to the isotropy of the material the response to $\bar{\sigma}_{22}^*$ must be:

$$(\sigma_a)_2 = -(\sigma_a)_1 \quad , \quad (\sigma_b)_2 = -(\sigma_b)_1$$

Hence, the response to $\bar{s}_{12}$ will be:

$$\sigma_a = (\sigma_a)_1 + (\sigma_a)_2 = (\sigma_a)_1 - (\sigma_a)_1 = 0$$

$$\sigma_b = (\sigma_b)_1 + (\sigma_b)_2 = (\sigma_b)_1 - (\sigma_b)_1 = 0$$

Secondly let us turn our attention to s_{ija}, s_{ijb}. The same macrostress $\bar{s}_{12}$ and the same rotation of axis will be considered. Let us investigate the possibility that $\bar{s}_{12}$ causes non-vanishing values of e. g. s_{13a}, s_{13b}. Any of these two components could be expressed as a set of two normal stress components in a rotated system of axis X_1^{**}, X_2, X_3^{**}, with X_2 the axis of rotation and with the angle of rotation 45°:

$$\sigma_{11a}^{**} = s_{13a} \quad , \quad \sigma_{33a}^{**} = -s_{13a}$$

$$\sigma_{11b}^{**} = s_{13b} \quad , \quad \sigma_{33b}^{**} = - s_{13b}$$

From the point of view of normal stressing the plus and minus sense of the axes indicating the direction of stressing is irrelevant and therefore the angles between X_1^* and X_1^{**}, X_1^* and X_3^{**} , X_2^* and X_1^{**}, X_2^* and X_3^{**} are the same /60°/. Accordingly, if the response to $\bar{\sigma}_{11}^*$ is $(s_{13a})_1$, i.e.:

$$(\sigma_{11a}^{**})_1 = (s_{13a})_1$$

the respective response to $\bar{\sigma}_{22}^*$ will be:

$$(s_{13a})_2 = (\sigma_{11a}^{**})_2 = -(\sigma_{11a}^{**})_1 = -(s_{13a})_1$$

and the respective response to $\bar{s}_{12}$:

$$s_{13a} = (s_{13a})_1 + (s_{13a})_2 = (s_{13a})_1 - (s_{13a})_1 = 0$$

Similarly we can proceed with other components and show that

$$s_{ija} = s_{ijb} = 0 \qquad \text{for} \qquad ij \neq 12 \; .$$

q. e. d.

The mesoscopic elastic response to the isotropic macroscopic stress component $\bar{\sigma}$ is only isotropic, i.e. σ_a , σ_b . This follows simply from the fact that the macroscopic loading as well as the material properties are isotropic and so must be the response. In details this could be shown again by replacing any of the deviatoric components s_{ija} or s_{ijb} by two normal components with opposite signs in a rotated coordinate system. By definition the signs must be opposite, but in the same time they must be equal, as the respective normal components cannot depend on the orientation due to the isotropy of the loading and due to the statistical isotropy of the material. Therefore they must be zero.

If the process is inelastic, the structure of the mathematical model given by the structural parameters is supposed to remain unchanged in the course of the described process, but apart from the immediate elastic response there can exist residual mesoscopic stresses due to the preceding history. These residual stresses form a self-equilibrated system and have no relation to the just acting macrostress.

The fluctuations of the mesoscopic stresses are described by the tensors σ'_{ijn} following in their shape that of the tensors σ_{ijn} , which is of course an approximate description.

I.2 Materials with Macroscopic Anisotropy in the Virgin State

In what follows we are going to describe materials that are composed of isotropic material constituents, but due to the geometry of their composition they are anisotropic already in the virgin state /defined by zero values of the macroscopic as well as microscopic stresses and strains/. The emphasizing of the virgin state is substantial, as the materials described in the preceding chapter can develop also anisotropy of some properties due to the deviatoric residual microstresses resulting from an inelastic process.

Another substantial difference is in the definition of the distribution functions and the respective structural parameters that are related here to the total components of the stress- and strain- tensors, not separately to their deviatoric and isotropic parts. Such an approach, if applied to the isotropic materials, would mean equality of the distribution functions and of the structural parameters for the deviatoric and the isotropic parts. This would simplify the model, but in some cases it means a serious restriction of its possibilities, as we have ascertained on concrete applications. Therefore, in the case of an isotropic material the distributions of the deviatoric and the isotropic parts are described separately. The possibility of doing so is clarified in paragraph I. 1. 14. However, no such possibility exists in the case of an anisotropic material. The reasoning used in I. 1. 14 cannot be used here.

The basic set of equations for the mathematical model is quite analogous to that of the isotropic case. Eqs. (1) , (2), (3) remain unchanged. Instead of eqs.(9) we write:

$$\varepsilon_{\alpha\beta}(\boldsymbol{X}, \boldsymbol{x}, t)_n = \bar{\varepsilon}_{\alpha\beta}(\boldsymbol{X}, t) + f^{\varepsilon}_{\alpha\beta}(\boldsymbol{X}, \boldsymbol{x}, t)_n \, \varepsilon'_{\alpha\beta n}(\boldsymbol{X}, t) \qquad (245)$$

$$\sigma_{\alpha\beta}(\boldsymbol{X}, \boldsymbol{x}, t)_n = \sigma^*_{\alpha\beta n}(\boldsymbol{X}, t) + f^{\sigma}_{\alpha\beta}(\boldsymbol{X}, \boldsymbol{x}, t)_n \, \sigma'_{\alpha\beta n}(\boldsymbol{X}, t)$$

with

$$\langle f^{\varepsilon}_{\alpha\beta} \rangle_n = \langle f^{\sigma}_{\alpha\beta} \rangle_n = 1 \qquad (246)$$

The averaging procedure leads to equations that are analogous to (12) :

$$\varepsilon'_{\alpha\beta n} = \varepsilon_{\alpha\beta n} - \bar{\varepsilon}_{\alpha\beta}$$

$$\sigma'_{\alpha\beta n} = \sigma_{\alpha\beta n} - \sigma^*_{\alpha\beta n} \qquad (247)$$

The fundamental theorem of the approach /I.1.2/ leads to the analogues of eqs. (14) - the model description of the stress- and strain- distributions:

$$\varepsilon_{\alpha\beta}(X,x,t)_n = \bar{\varepsilon}_{\alpha\beta}(X,t) + f^{c}_{\alpha\beta}(x)_n\, \varepsilon'_{\alpha\beta n}(X,t) \tag{248}$$

$$\sigma_{\alpha\beta}(X,x,t)_n = \sigma^{*}_{\alpha\beta n}(X,t) + f^{\sigma}_{\alpha\beta}(x)_n\, \sigma'_{\alpha\beta n}(X,t)$$

/no summation on α, β/.

The specific stress power in a unit volume of the composite is expressed by the respective analogue of eq. (17):

$$\overline{\sigma_{ij}\,\dot{\varepsilon}_{ij}} = \sum_{n=1}^{N} v_n \left(\sigma_{ijn}\,\dot{\varepsilon}_{ijn} + \Xi_{ijn}\,\dot{\varepsilon}'_{ijn}\right) \tag{249}$$

where

$$\Xi_{\alpha\beta n} = \frac{\sigma'_{\alpha\beta n}}{\eta_{\alpha\beta n}} \tag{250}$$

$$\eta_{\alpha\beta n} = \left\langle \left[f^{\sigma}_{\alpha\beta}(x)_n - 1\right]\left[f^{c}_{\alpha\beta}(x)_n - 1\right]\right\rangle_n = \left\langle f^{\sigma}_{\alpha\beta}(x)_n\, f^{c}_{\alpha\beta}(x)_n\right\rangle_n - 1 \tag{251}$$

Instead of eq. (28), we can quite similarly deduce:

$$\sigma_{\alpha\beta n} - \sigma_{\alpha\beta m} + \frac{\sigma'_{\alpha\beta n}}{\eta_{\alpha\beta n}} - \frac{\sigma'_{\alpha\beta m}}{\eta_{\alpha\beta m}} = 0 \tag{252}$$

Equations (30) and (32) remain unchanged and thus, the basic set of equations forming the starting point for the user of the model is represented by eqs. (1), (2), (30), (32), (247) and (252), i.e. equally as in the isotropic case by 4N + 1 equations for 4N + 1 unknown rates $\dot{\sigma}_{ijn}$, $\dot{\sigma}'_{ijn}$, $\dot{\varepsilon}_{ijn}$, $\dot{\varepsilon}'_{ijn}$, $\dot{\bar{\varepsilon}}_{ij}$.

For the dual model /B-model/ in eqs. (247) and (252) stresses and strains are interchanged and symbols $\eta_{\alpha\beta n}$, $\eta_{\alpha\beta m}$ are replaced by $\chi_{\alpha\beta n}$, $\chi_{\alpha\beta m}$.

Similarly as in I.1.8 it can easily be shown that for a two-phase material it holds:

$0 < \eta_{\alpha\beta n} < \infty$, $0 < \eta_{\alpha\beta m} < \infty$ describes compact infrastructures of both material constituents

$0 < \chi_{\alpha\beta n} < \infty$, $0 < \chi_{\alpha\beta m} < \infty$ describes loose infrastructures of both material constituents

$\eta_{\alpha\beta n} = \infty, 0 < \eta_{\alpha\beta m} < \infty$

or $\chi_{\alpha\beta n} = \infty, 0 < \chi_{\alpha\beta m} < \infty$ describes loose infrastructure of the n- constituent and compact infrastructure of the m- constituent

$\eta_{\alpha\beta n} = \infty, \eta_{\alpha\beta m} = \infty$

or $\chi_{\alpha\beta n} = 0, \chi_{\alpha\beta m} = 0$ describes the homogeneous stress model /Reuss' solution/

$\eta_{\alpha\beta n} = 0, \eta_{\alpha\beta m} = 0$

or $\chi_{\alpha\beta n} = \infty, \chi_{\alpha\beta m} = \infty$ describes the homogeneous strain model /Voigt's solution/

I.2.1 Transversely Isotropic Materials

Contrary to the isotropic case the structural parameters $\eta_{\alpha\beta n}$ or $\chi_{\alpha\beta n}$ are not independent of the tensor components to which they relate. This complicates the model substantially and it seems that to remain in the framework of a reasonably simple operative model we must restrict ourselves to the case of transverse isotropy. Then - with the X_1 coordinate axis taken for the axis of symmetry - equation (252) can be rewritten in the following form /for proof see Appendix 6/:

$$\sigma_{ijn} - \sigma_{ijm} + \frac{\sigma'_{ijn}}{[\eta_n]} - \frac{\sigma'_{ijm}}{[\eta_m]} = 0 \qquad (253)$$

or - for the B-model:

$$\varepsilon_{ijn} - \varepsilon_{ijm} + \frac{\varepsilon'_{ijn}}{[\chi_n]} - \frac{\varepsilon'_{ijm}}{[\chi_m]} = 0 \qquad (254)$$

where

$$[\eta_n] = \eta_n, \quad [\chi_n] = \chi_n \quad \text{for} \quad ij = 11 \qquad (255)$$

$$[\eta_n] = \eta_n^{\perp}, \quad [\chi_n] = \chi_n^{\perp} \quad \text{for} \quad ij = 22, 23, 33$$

$$[\eta_n] = \eta_n^{\times}, \quad [\chi_n] = \chi_n^{\times} \quad \text{for} \quad ij = 12, 13$$

I.2.2 An Important Example of a Transversely Isotropic Two-Phase Model

A number of concrete applications can be derived as special cases of the following model that is a transversely isotropic analogue to that described in I.1.9. Equations (38), (39), $(40)_1$, $(40)_2$, (42), (43), (44) and (45) remain unchanged, instead of eqs. (41) we have eqs. (253). In the case of the dual B-model instead of eqs. (40) and (253) we make use of eqs. (56) and (254).

To formulate the macroscopic constitutive equation of the A-model, we can easily deduce from eqs. (39), (44), (42) and (38):

$$\dot{\bar{\varepsilon}}_{ij} = \mu_e \dot{\bar{s}}_{ij} + v_n(\mu_n - \mu_e)\dot{s}_{ijn} + v_n s_{ijn}\dot{h} + \delta_{ij}\big[\varrho_e \dot{\bar{\sigma}} + \qquad (256)$$
$$+ v_n(\varrho_n - \varrho_e)\dot{\sigma}_n + v_n \sigma_n \dot{\varrho}_n + (v_e \alpha_e + v_n \alpha_n)\dot{T} + v_n \dot{\tau}\big]$$

This equation comprises the rates of macroscopic stress and apart from them the rates $\dot{s}_{ijn}$, $\dot{\sigma}_n$, which must be also expressed in terms of the macro-stress-rates. To this purpose we give at first the formulae for $\dot{\sigma}'_{ije}$ and $\dot{\sigma}'_{ijn}$ in terms of $\dot{\bar{\sigma}}_{ij}$, $\dot{\sigma}_{ijn}$ using equations (45), (40), (44), (39), (42), (38), (43):

$$\dot{\sigma}'_{ije} = \frac{v_n}{v_e}\left(\dot{\bar{\sigma}}_{ij} - \frac{v_e \mu_n + v_n \mu_e}{\mu_e}\dot{\sigma}_{ijn}\right) - \frac{v_n}{\mu_e} s_{ijn}\dot{h} - \qquad (257)$$
$$- \delta_{ij}\frac{v_n}{\varrho_e}\left[\sigma_n \dot{\varrho}_n + (\alpha_n - \alpha_e)\dot{T} + \dot{\tau}\right] + \delta_{ij} v_n\left(\frac{\mu_n}{\mu_e} - \frac{\varrho_n}{\varrho_e}\right)\dot{\sigma}_n$$

$$\dot{\sigma}'_{ijn} = \frac{1}{\mu_n}\left[-\mu_e \dot{\bar{\sigma}}_{ij} + (v_e \mu_n + v_n \mu_e)\dot{\sigma}_{ijn} + (v_e s_{ijn} - s'_{ijn})\dot{h}\right] + \qquad (258)$$
$$+ \delta_{ij}\frac{1}{\varrho_n}\left[(v_e \sigma_n - \sigma'_n)\dot{\varrho}_n + v_e(\alpha_n - \alpha_e)\dot{T} + v_e \dot{\tau}\right] +$$
$$+ \delta_{ij}\left(\frac{\mu_e}{\mu_n} - \frac{\varrho_e}{\varrho_n}\right)(\dot{\bar{\sigma}} - v_n \dot{\sigma}_n)$$

In relation to the inelastic deformations that are the main phenomenon to be described the influence of the differing elastic Poisson's ratios is often negligeable and therefore in what follows we are going to give the formulae only for the special case:

$$\nu_e = \nu_n = \nu$$

With this equality valid the last addends on the right-hand sides of eqs. (257) and (258) are vanishing. Eqs. (257) and (258) modified in this way are then used in eq. (253) /index m replaced by e, derivatives instead of stresses/. The deduction of the expression for $\dot{\sigma}_{ijn}$ is then straightforward:

$$\dot{\sigma}_{ijn} = [M]\,\dot{\bar{\sigma}}_{ij} - \left([N]s_{ijn} - [N']s'_{ijn}\right)\dot{h} - \tag{259}$$

$$- \delta_{ij}\frac{\mu_e}{\varrho_e}\left\{\left([N]\sigma_n - [N']\sigma'_n\right)\dot{\varrho}_n + [N](\alpha_n - \alpha_e)\dot{T} + [N]\dot{\tau}\right\}$$

with

$$[M] = \left\{\mu_e\mu_n[\eta_e][\eta_n] + \mu_e\left(v_e\mu_e[\eta_e] + v_n\mu_n[\eta_n]\right)\right\}/[R] \tag{260}$$

$$[N] = \frac{1-[M]}{\mu_n - \mu_e} = v_e\left(v_e\mu_e[\eta_e] + v_n\mu_n[\eta_n]\right)/[R]$$

$$[N'] = v_e\mu_e[\eta_e]/[R]$$

$$[R] = \mu_e\mu_n[\eta_e][\eta_n] + (v_e\mu_n + v_n\mu_e)(v_e\mu_e[\eta_e] + v_n\mu_n[\eta_n])$$

where

$$[M] = M(\eta_e, \eta_n) \quad \text{for} \quad ij = 11 \tag{261}$$

$$[M] = M^{\perp}(\eta_e^{\perp}, \eta_n^{\perp}) \quad \text{for} \quad ij = 22, 23, 33$$

$$[M] = M^{\times}(\eta_e^{\times}, \eta_n^{\times}) \quad \text{for} \quad ij = 12, 13$$

and similar meaning have the symbols $[N]$, $[N']$, $[R]$ /the formulae are quite analogous to eqs. $(49)_1$, $(49)_3$, $(49)_4$, $(49)_5$ /.

In the case of the B-model eq. (256) remains unchanged, from eqs. (45) , (56) , (43) and (38) we arrive at :

$$\dot{\varepsilon}'_{ijn} = \mu_n(\dot{s}_{ijn} - \dot{\bar{s}}_{ij}) + \dot{h}(s_{ijn} - \bar{s}_{ij}) + \delta_{ij}\left[\varrho_n(\dot{\sigma}_n - \dot{\bar{\sigma}}) + \dot{\varrho}_n(\sigma_n - \bar{\sigma})\right] \tag{262}$$

$$\dot{\varepsilon}'_{ije} = \frac{v_n}{v_e}\left[\mu_e(\dot{\bar{s}}_{ij} - \dot{s}_{ijn}) + \delta_{ij}\varrho_e(\dot{\bar{\sigma}} - \dot{\sigma}_n)\right]$$

and from eqs. (254) , (39) , (44) , (42) , (38) and (262) we finally get :

$$\dot{s}_{ijn} + \delta_{ij}\frac{\varrho_e}{\mu_e}\dot{\sigma}_n = [P]\left(\dot{\bar{s}}_{ij} + \delta_{ij}\frac{\varrho_e}{\mu_e}\dot{\bar{\sigma}}\right) - \tag{263}$$

$$- [Q]\left\{\left[(1+[\chi_n])s_{ijn} - \bar{s}_{ij}\right]\dot{h} + \delta_{ij}\left[(1+[\chi_n])\sigma_n - \right.\right.$$

$$\left.\left. - \bar{\sigma}\right]\dot{\varrho}_n + \delta_{ij}\left[(\alpha_n - \alpha_e)\dot{T} + \dot{\tau}\right][\chi_n]\right\}$$

with

$$[P] = \left\{ \mu_e [\chi_e][\chi_n] + v_e \mu_n [\chi_e] + v_n \mu_e [\chi_n] \right\} / [R] \tag{264}$$

$$[Q] = v_e [\chi_e] / [R]$$

$$[R] = (v_e \mu_n + v_n \mu_e) [\chi_e][\chi_n] + v_e \mu_n [\chi_e] + v_n \mu_e [\chi_n]$$

The meaning of the square brackets is analogous as in the case of $[M]$.

It is important to note that due to the equality of Poisson's ratios it holds:

$$v_n (\mu_n - \mu_e) \dot{s}_{ijn} + \delta_{ij} v_n (\rho_n - \rho_e) \dot{\sigma}_n =$$
$$= v_n (\mu_n - \mu_e) \left(\dot{s}_{ijn} + \delta_{ij} \frac{\rho_e}{\mu_e} \dot{\sigma}_n \right)$$

which makes the use of eq. (263) in (256) very easy.

In this way we have presented the macroscopic constitutive equations for the A- and the B- model with the tensorial internal variables $\sigma_{ijn}, \sigma'_{ijn}$ and their evolution equations.

Similarly as in the case of an isotropic material it is possible to deduce from the A-model as well as from the B-model their special variants for inclusions in a matrix, for homogeneous stress and homogeneous strain.

Let us clarify the above relations on the case of an elastic isothermal process. From eqs. (256) and (259) we get / A-model/:

$$\dot{\bar{\varepsilon}}_{ij} = \left[\mu_e + v_n(\mu_n - \mu_e)[M]\right] \dot{\bar{s}}_{ij} + \delta_{ij}\left[\rho_e + v_n(\mu_n - \mu_e)[M]\right] \dot{\bar{\sigma}} + \tag{265}$$
$$+ \delta_{ij} v_n (\rho_n - \rho_e + \mu_e - \mu_n) \frac{1}{3} \left[M \dot{\bar{\sigma}}_{11} + M^{\perp} (\dot{\bar{\sigma}}_{22} + \dot{\bar{\sigma}}_{33}) \right]$$

and from eq. (256) and (263) /B-model/:

$$\dot{\bar{\varepsilon}}_{ij} = \left[\mu_e + v_n(\mu_n - \mu_e)[P]\right] \dot{\bar{s}}_{ij} + \delta_{ij}\left[\rho_e + \right. \tag{266}$$
$$\left. + v_n(\mu_n - \mu_e) \frac{\rho_e}{\mu_e} [P]\right] \dot{\bar{\sigma}}$$

If we denote

$$[\bar{\mu}] = \bar{\mu} = \frac{\bar{e}_{11}}{\bar{s}_{11}} = \frac{\bar{\varepsilon}_{11} - \bar{\varepsilon}_{22}}{\bar{\sigma}_{11}} \quad \text{for uniaxial loading } \bar{\sigma}_{11} \tag{267}$$

$$= \bar{\mu}^{\perp} = \frac{\bar{e}_{22}}{\bar{s}_{22}} = \frac{2\bar{\varepsilon}_{22} - \bar{\varepsilon}_{11} - \bar{\varepsilon}_{33}}{2\,\bar{\sigma}_{22}} \quad \text{—— " ——} \quad \bar{\sigma}_{22}$$

$$=\bar{\mu}^{\perp}=\frac{\bar{e}_{23}}{\bar{s}_{23}}=\frac{\bar{\varepsilon}_{23}}{\bar{\sigma}_{23}} \qquad \text{—— " ——} \qquad \bar{\sigma}_{23}$$

$$=\bar{\mu}^{\times}=\frac{\bar{e}_{12}}{\bar{s}_{12}}=\frac{\bar{\varepsilon}_{12}}{\bar{\sigma}_{12}} \qquad \text{—— " ——} \qquad \bar{\sigma}_{12}$$

we can easily deduce from the preceding equations:

for the A-model:

$$[\bar{\mu}]=\mu_e+v_n(\mu_n-\mu_e)[M]= \tag{268}$$

$$=\frac{(v_e\mu_e+v_n\mu_n)[\eta_e][\eta_n]+v_e\mu_e[\eta_e]+v_n\mu_n[\eta_n]}{\mu_e\mu_n[\eta_e][\eta_n]+(v_e\mu_n+v_n\mu_e)(v_e\mu_e[\eta_e]+v_n\mu_n[\eta_n])}\mu_e\mu_n$$

and for the B-model:

$$[\bar{\mu}]=\mu_e+v_n(\mu_n-\mu_e)[P] \tag{269}$$

$$=\frac{\mu_e\mu_n[\chi_e][\chi_n]+(v_e\mu_e+v_n\mu_n)(v_e\mu_n[\chi_e]+v_n\mu_e[\chi_n])}{(v_e\mu_n+v_n\mu_e)[\chi_e][\chi_n]+v_e\mu_n[\chi_e]+v_n\mu_e[\chi_n]}$$

The above expressions are quite similar to those defined by eqs. (48) and (60) for an isotropic material.

If one of the compliances - e.g. μ_n - is infinite/vanishing shear modulus/. we get from eq. (268) :

$$[\bar{\mu}]=\frac{1+[\eta_e]}{v_e}\mu_e \tag{270}$$

whereas for the B-model the resulting values of $[\bar{\mu}]$ are infinite /vanishing macroscopic shear moduli/ - in accordance with the loose infrastructure that is described by this model.

Due to the assumed equality of Poisson's ratios similar relations follow for the macroscopic E-moduli in the case of $\bar{\sigma}_{11}$ loading and $\bar{\sigma}_{22}$ - loading:

A-model:

$$[\bar{E}]=\frac{E_e E_n}{E_n+v_n[M](E_e-E_n)}= \tag{271}$$

$$=\frac{[\eta_e][\eta_n]E_eE_n+(v_eE_e+v_nE_n)(v_e[\eta_e]E_n+v_n[\eta_n]E_e)}{[\eta_e][\eta_n](v_eE_n+v_nE_e)+v_e[\eta_e]E_n+v_n[\eta_n]E_e}$$

where for the loading $\bar{\sigma}_{11}$, $[\bar{E}]$, $[\eta_e]$ and $[\eta_n]$ have the meaning of $\bar{E}$, η_e, η_n, and for $\bar{\sigma}_{22}$ the meaning of $\bar{E}^{\perp}$, $\eta_e^{\perp}$, $\eta_n^{\perp}$.

B-model:

$$[\bar{E}] = \frac{E_e E_n}{E_n + v_n [P](E_e - E_n)} = \tag{272}$$

$$= \frac{[\chi_e][\chi_n](v_e E_e + v_n E_n) + v_e[\chi_e]E_e + v_n[\chi_n]E_n}{[\chi_e][\chi_n]E_e E_n + (v_e E_n + v_n E_e)(v_e[\chi_e]E_e + v_n[\chi_n]E_n)} E_e E_n$$

If one of the moduli - e.g. E_n - is vanishing, we get from eq. (271) for the A-model:

$$[\bar{E}] = \frac{v_e}{1 + [\eta_e]} E_e \tag{273}$$

For the B-model, the resulting macroscopic moduli following from eq. (272) are vanishing.

According to our general considerations the model corresponding to inclusions e.g. of the e-material in the matrix of the n-material is described as a special case of the A-model with infinite parameters $[\eta_e]$ or as a special case of the B-model with infinite parameters $[\chi_e]$. From eqs. (271) and (272) we get for such cases:

$$[\bar{E}] = \frac{[\eta_n]E_e + v_e(v_e E_e + v_n E_n)}{[\eta_n](v_e E_n + v_n E_e) + v_e E_n} E_n \qquad \text{(A)} \tag{274}$$

$$[\bar{E}] = \frac{[\chi_n](v_e E_e + v_n E_n) + v_e E_e}{[\chi_n] E_n + v_e(v_e E_n + v_n E_e)} E_n \qquad \text{(B)} \tag{275}$$

These two expressions are equivalent, connected by the relation

$$[\eta_n] = \frac{v_e^2}{[\chi_n]} \tag{276}$$

similarly as in section I.1.9.3.

I.2.3. The Solution to the Identification Problem Based on the Flow-Curve for a Transversely Isotropic Material

Let us assume that the material in question is composed of two material constituents with compact infrastructures, one of the constituents being purely elastic, the other with elastic response and linear viscosity. The process under study is isothermal. The adequate mathematical model is the A-model with

$$\dot{h} = \frac{1}{2H}, \quad \dot{g}_n = 0, \quad \dot{T} = 0, \quad \dot{\tau} = 0 \tag{277}$$

in eqs. (256), (257), (258) and (259).

For the solution to the identification problem let us start with the simplest case, where we know all the constants except for the structural parameters. The first piece of information are the experimentally determined macroscopic moduli. If $\bar{E}$ and $\bar{E}^{\perp}$ are known, eq. (271) yields two equations for four unknown parameters η_e, $\eta_e^{\perp}$, η_n, $\eta_n^{\perp}$.

Furthermore, let us have at our disposal the flow curves under constant loads $\bar{\sigma}_{11}$ and $\bar{\sigma}_{22}$ with measurements of deformations in the longitudinal as well as in the tracesverse directions. Let us denote by $\bar{\varepsilon}_{11}^{(as)}, \bar{\varepsilon}_{22}^{(as)}, \bar{\varepsilon}_{33}^{(as)}$ the values of deformations that are asymptotically approached with time passing. In the course of the deformation process the deviatoric stresses s_{ijn}, s'_{ijn} asymptotically approach zero values and the deformation state approaches an elastic state with μ_n infinite /shear modulus of the n-material vanishing/. Then - according to (267)- we can write:

$$\bar{\mu}^{(as)} = \frac{\bar{\varepsilon}_{11}^{(as)} - \bar{\varepsilon}_{22}^{(as)}}{\bar{\sigma}_{11}} \tag{278}$$

$$\bar{\mu}^{\perp(as)} = \frac{2\bar{\varepsilon}_{22}^{(as)} - \bar{\varepsilon}_{11}^{(as)} - \bar{\varepsilon}_{33}^{(as)}}{2\bar{\sigma}_{22}} \tag{279}$$

and using these experimentally determined values in eq. (270) we have two equations for η_e, $\eta_e^{\perp}$. Hence, together we have four equations for four structural parameters η_e, η_n, $\eta_e^{\perp}$, $\eta_n^{\perp}$. The remaining structural parameters η_e^{x}, η_n^{x} can be determined quite similarly - if necessary - from the flow-curve under $\bar{\sigma}_{12}$.

So far we have assumed only deviatoric viscosity. There exist cases, where the n-material contains interconnected pores with liquid and under hydrostatic pressure or tension this liquid can leak out or inside. On the mesoscale this phenomenon appears as volume viscosity. In our model it can easily be described if $\dot{\varrho}_n$ is replaced by a constant - the coefficient of volume viscosity. Such model has meaning especially for biological materials - cf. e.g. Kafka and Jírová [37]. Then, if a sample of such a material is loaded by a fixed value of $\bar{\sigma}_{11}$ or $\bar{\sigma}_{22}$, the stresses relaxed in the asymptotic state are not only the deviatoric parts, but also the isotropic parts. Hence, we can use formula (273) for the determination of the structural parameters η_e, $\eta_e^{\perp}$. The advantage is that we do not need the measurements of the transverse deformations as in the case we use eq. (270). Unfortunatelly these two different cases /possibility of use of eq. (270) or (273) were not clearly distinguished in Kafka [36] and [37].

In reality, the knowledge of all the material constants except for the structural parameters is not very probable and therefore we must seek other sources of information. If we introduce the second-order derivatives, we are able to eliminate the internal variables from the macroscopic constitutive equation. We are going to give the formulae for the special case of uniaxial loading $\bar{\sigma}_{11}$ with $\nu_e = \nu_v = 0.5$, i.e. $\varrho_e = \varrho_v = 0$. The elastic change of volume is thus neglected. The resulting rheological equation, derived from eqs. (256), (259) and (258) is:

$$\ddot{\bar{\varepsilon}}_{11} + A\dot{\bar{\varepsilon}}_{11} + B\bar{\varepsilon}_{11} = C\ddot{\bar{s}}_{11} + D\dot{\bar{s}}_{11} + F\bar{s}_{11} \tag{280}$$

where

$$A = [1 + \mu_n \hat{N} - (v_e \mu_n + v_n \mu_e)\hat{N}']/2H\mu_n \tag{281}$$

$$B = (\hat{N} - v_e \hat{N}')/4H^2\mu_n$$

$$C = \mu_e + v_n(\mu_n - \mu_e)M$$

$$D = [\mu_e + v_n(2\mu_n - \mu_e)M + \mu_e\mu_n(\hat{N} - \hat{N}')]/2H\mu_n$$

$$F = [v_n M + \mu_e(\hat{N} - \hat{N}')]/4H^2\mu_n$$

$$\hat{N} = (2N + N^{\perp})/3 \tag{282}$$

$$\hat{N}' = (2N' + N'^{\perp})/3$$

For a flow curve corresponding to $\bar{\sigma}_{11} = const.$ we have $\ddot{\bar{s}}_{11} = \dot{\bar{s}}_{11} = 0$ and eq. (280) simplifies and comprises only A, B and F. This differential equation can be identified with the experimental flow-curve and the values of A, B and F determined. The simplest way of doing so is to determine the values of $\ddot{\bar{\varepsilon}}_{11}, \dot{\bar{\varepsilon}}_{11}, \bar{\varepsilon}_{11}, \bar{s}_{11}$ in three points, which leads to three linear equations for A, B, F. Supposing that M is known from eq. (271) it is easy to calculate $H, \hat{N}$ and $\hat{N}'$. The concrete procedure depends upon what is known a priori. The main object of interest is the value of H, which is very difficult to determine or to estimate in another way.

The whole procedure of the solution to the identification problem and especially of the determination of the structural parameters and of H simplifies substantially if the structure of the material in question is formed by elastic inclusions in a visco elastic matrix. The structural parameters $[\eta_n]$ can be determined from the macroscopic moduli using eq. (274). For infinite values of $[\eta_e]$ /descriptive of the elastic inclusions/ B is zero and for growing time $\ddot{\bar{\varepsilon}}_{11}$ becomes vanishing and $\dot{\bar{\varepsilon}}_{11}$ approaches a constant value that follows from eqs. (280) and (281):

$$(\dot{\bar{\varepsilon}}_{11})_{t\to\infty} = \frac{(F)_{[\eta_e]=\infty}}{(A)_{[\eta_e]=\infty}}\bar{s}_{11} = \frac{v_n}{2H}\cdot\frac{\Omega_1}{\Omega_2}\bar{s}_{11} \tag{283}$$

with

$$\Omega_1 = 3\mu_n \eta_n \eta_n^{\perp} + v_e(3v_e\mu_n + 3v_n\mu_e - \mu_e)\eta_n + v_e\mu_e\eta_n^{\perp} \qquad (284)$$

$$\Omega_2 = 3\mu_n \eta_n \eta_n^{\perp} + v_e(3v_e\mu_n + 2v_n\mu_e)\eta_n + v_e(3v_e\mu_n + v_n\mu_e)\eta_n^{\perp} + 3v_e^3(v_e\mu_n + v_n\mu_e)$$

Hence, from the asymptotic direction of the flow-curve we can easily determine H.

In the case of an isotropic material $\eta_n = \eta_n^{\perp}$ and eq. (283) simplifies substantially:

$$(\dot{\varepsilon}_{11})_{t\to\infty} = \frac{v_n}{2H} \cdot \frac{\eta_n}{\eta_n + v_e^2}\,\bar{s}_{11} \qquad (285)$$

i.e. to an expression that can be arrived at from eq. (192) for deviatoric strain and stress with η_e infinite.

I.2.4 Transversely Isotropic Materials with Unidirectional Continuous Fibers

A special and important example of transversely isotropic materials are composites reinforced with unidirectional continuous fibers that are randomly distributed.

Similarly as in I.2.2 we assume validity of eqs. (38), (39), (40), (42), (43), (44) and (45), but instead of eqs. (253) we have:

$$\varepsilon_{11e} = \varepsilon_{11n} = \bar{\varepsilon}_{11} \qquad (286)$$

$$\sigma_{ijn} - \sigma_{ije} + \frac{\sigma'_{ijn}}{[\eta_n]} = 0 \qquad (287)$$

where

$$[\eta_n] = \eta_n^{\perp} \quad \text{for} \quad ij = 22, 23, 33$$
$$[\eta_n] = \eta_n^{\times} \quad \text{for} \quad ij = 12, 13 \qquad (288)$$

The meaning of eq. (286) is clear without comments, eqs. (287) follow from (253) if $[\eta_e]$ are infinite, which corresponds to fibers that represent a special case of inclusions.

Eqs. (256) to (258) are valid for all indices, as the relations from which they were derived hold true whithout any limitation. Again we assume $\nu_e = \nu_n = \nu$.

For $ij \neq 11$ eq. (259) is also valid, but the definitions of $[M]$, $[N]$, $[N']$ are different due to the infinite values of $[\eta_e]$:

$$[M] = (\mu_n[\eta_n] + v_e\mu_e)/[R] \qquad (289)$$

$$[N] = v_e^2 / [R]$$

$$[N'] = v_e / [R]$$

where $[R] = \mu_n [\eta_n] + v_e (v_e \mu_n + v_n \mu_e)$

$$[M] = M^{\perp}(\eta_n^{\perp}) \quad \text{for} \quad ij = 22, 23, 33$$

$$[M] = M^{*}(\eta^{*}) \quad \text{for} \quad ij = 12, 13$$

and similarly for $[N], [N'], [R]$.

For $ij = 11$ eq. (259) is not valid, instead eq. (286) with (42) and (44) leads to:

$$\dot{s}_{11n} + \frac{\varrho_e}{\mu_e} \dot{\sigma}_n = \frac{1}{v_e \mu_n + v_n \mu_e} \Big[\mu_e \dot{\bar{s}}_{11} + \varrho_e \dot{\bar{\sigma}} - v_e s_{11n} \dot{h} - v_e \sigma_n \dot{\varrho}_n - v_e (\alpha_n - \alpha_e) \dot{T} - v_e \dot{\tau} \Big] \tag{290}$$

This expression can directly be used in eq. (256), as for the assumed equality of Poisson's ratios it holds:

$$v_n (\mu_n - \mu_e) \dot{s}_{11n} + v_n (\varrho_n - \varrho_e) \dot{\sigma}_n = v_n (\mu_n - \mu_e) \left(\dot{s}_{11n} + \frac{\varrho_e}{\mu_e} \dot{\sigma}_n \right)$$

This completes the macroscopic constitutive equation.

I.2.4.1 Macroscopic Yield Condition of a Material with Unidirectional Continuous Fibers

Let us assume that that the fibers /e-material/ remain in the elastic state whereas the matrix /n-material/ is perfectly elastic--plastic with the simplest form of the Mises criterion:

$$s_{ijn} s_{ijn} = 2k^2 \tag{291}$$

We are going to deduce the respective macroscopic yield condition. Before loading the composite is assumed to be without internal stresses and up to the state where eq. (291) is fulfilled the deformation process is elastic. Therefore, it holds according to eqs. (290), (259) and (289):

for $ij = 11$:

$$s_{11n} + \frac{\varrho_e}{\mu_e} \sigma_n = \frac{\mu_e \bar{s}_{11} + \varrho_e \bar{\sigma}}{v_e \mu_n + v_n \mu_e} \tag{292}$$

for $ij = 22, 33, 23, 12, 13$:

$$\sigma_{ijn} = s_{ijn} + \delta_{ij} \sigma_n = [M] \bar{\sigma}_{ij} \tag{293}$$

where $[M]$ is given by eq. $(289)_1$.

From eqs. (292) and (293) we easily express all the deviatoric components s_{ijn} in terms of $\bar{\sigma}_{ij}$ and using these expressions in eq. (291)

we arrive at:

$$\bar{s}_{11}^2 \frac{6\hat{\mu}^2 + (\hat{\varrho}^2 - 2\hat{\varrho} - 2)M^{\perp 2} + 6\hat{\mu}\hat{\varrho}M^{\perp}}{(2+\hat{\varrho})^2} + \tag{294}$$

$$+(\bar{s}_{22}^2 + \bar{s}_{33}^2 + 2\bar{s}_{23}^2)M^{\perp 2} + 2(\bar{s}_{12}^2 + \bar{s}_{13}^2)M^{\times 2} +$$

$$+\bar{\sigma}^2 6\left(\hat{\varrho}\,\frac{\hat{\mu} - M^{\perp}}{2+\hat{\varrho}}\right)^2 + \bar{s}_{11}\bar{\sigma}\,6\hat{\varrho}\,\frac{(\hat{\mu} - M^{\perp})(2\hat{\mu} + \hat{\varrho}M^{\perp})}{(2+\hat{\varrho})^2} = 2k^2$$

where we have newly introduced dimensionless constants:

$$\hat{\mu} = \frac{\mu_e}{v_e\mu_n + v_n\mu_e} = \frac{\varrho_e}{v_e\varrho_n + v_n\varrho_e} = \frac{E_n}{v_eE_e + v_nE_n} \tag{295}$$

$$\hat{\varrho} = \frac{\varrho_e}{\mu_e} = \frac{\varrho_n}{\mu_n} = \frac{1-2\mu}{1+\nu} \tag{296}$$

Eq. (294) with definitions (295), (296) represents the macroscopic transversely isotropic yield criterion for the special case of equal Poisson's ratios $\nu_e = \nu_n = \nu$. In the more general case of unequal Poisson's ratios the overall form with regard to the parts of $\bar{\sigma}_{ij}$ remains the same, only the coefficients are more complicated /cf. Kafka [30], page 692/. Thus, the resulting criterion depends on the isotropic part of the macrostress $\bar{\sigma}$, or - in the other words - on the hydrostatic pressure. This is in full agreement with the findings of other authors proceeding with other methods - e.g. of T.H.Lin, D.Salinas and Y.M.Ito [46].

However, if both material constituents are incompressible $/\nu_e = \nu_n = 0.5/$,

$$\hat{\varrho} = 0,$$

eq. (294) simplifies substantially and is independent of the isotropic part $\bar{\sigma}$. The equation that results in this way is - with regard to the parts of $\bar{\sigma}_{ij}$ - equivalent to the criterion proposed by R.Hill [16]. p. 319 . Only the form presented by R.Hill is a bit different and the coefficients appearing in his form are simply constants, with no relation to the microstructure.

I.2.4.2 The Solution to the identification problem based on the stress-strain diagram for a material with unidirectional continuous fibers.

Let us assume that the material under study is composed of a perfectly elastic-plastic matrix and unidirectional elastic fibers. Elastic Poisson's ratios of both the material constituents are supposed equal - $\nu_e = \nu_n = \nu$. The yield condition fo the matrix /n-material/ is assumed in the form given by eq. (291).

For a uniaxial loading $\bar{\sigma}_{11}$ we easily derive the respective formula for the elastic deformation $\bar{\varepsilon}_{11}$ from eqs. (256) and (290). These equations lead to:

$$\bar{\varepsilon}_{11} = \frac{\bar{\sigma}_{11}}{v_e E_e + v_n E_n} \tag{297}$$

i.e.

$$\bar{E} = v_e E_e + v_n E_n \tag{298}$$

Thus, in the case of equal Poisson's ratios the macroscopic modulus in the direction of the fibers is given by the "rule of mixtures" and is independent of Poisson's ratio and of the structural parameters.

Another situation will turn out if the composite is loaded by $\bar{\sigma}_{22}$. Again we can start from eq. (256), but this time it is necessary to express $\dot{\sigma}_{ijn}$ from (259), where $M^{\perp}$ is defined by eq. (289). After a short handling we get:

$$\bar{E}^{\perp} = \frac{\bar{\sigma}_{22}}{\bar{\varepsilon}_{22}} = \frac{2\mu_e + \varrho_e}{(\mu_e + 2\varrho_e)\left[\mu_e + v_n M^{\perp}(\mu_n - \mu_e)\right] + \mu_n \dfrac{(\mu_e - \varrho_e)^2}{3(v_e \mu_n + v_n \mu_e)}} \tag{299}$$

In eq. (299) ϱ_n does not appear, as it was expressed in terms of ϱ_e, μ_e, μ_n:

$$\varrho_n = \varrho_e \frac{\mu_n}{\mu_e} \tag{300}$$

Let us mention that for a rigid matrix / $E_n = \infty$, $\mu_n = 0$ / we get $M^{\perp} = 1/v_n$ and $\bar{E}^{\perp} = \infty$.

For rigid fibers / $E_e = \infty$, $\mu_e = 0 = \varrho_e$ / we arrive at:

$$\bar{E}^{\perp} = \frac{v_e^2 + \eta_n^{\perp}}{v_n (1 - \nu^2)\, \eta_n^{\perp}} E_n \tag{301}$$

Here, the vanishing value of $\eta_n^{\perp}$ corresponds to Voigt's solution /assumption of homogeneous strain/ and leads of course to an infinite value of $\bar{E}^{\perp}$.

Furthermore, let us analyse the case of loading by $\bar{\sigma}_{12}$. Starting from eqs. (256), (259), (289) we easily derive for the elastic domain:

$$\bar{\mu}^{\times} = \frac{\bar{\varepsilon}_{12}}{\bar{\sigma}_{12}} = \mu_e + v_n (\mu_n - \mu_e) M^{\times} \tag{302}$$

with $M^{\times}$ defined by eq. $(289)_1$.

In the simplest case of the identification procedure we suppose that the only unknown parameters are $E_n, \eta_n^{\perp}, \eta_n^{\times}, k$. The input data are: the measured values of $\bar{E}$, $\bar{E}^{\perp}$, $\bar{\mu}^{\times}$ and e.g. of $\bar{\sigma}_{22}^{L}$. the last symbol meaning the plastic limit under a uniaxial loading $\bar{\sigma}_{22}$. Similarly as in I.1.10 $\bar{\sigma}_{22}^{L}$ is to be determined as the value corresponding to the intersection of the straight elastic part of the stress-strain diagram with the backward extrapolation of the smooth plastic part with decreasing curvature. The following procedure is straightforward: from eq. (298) we determine E_n, from (299) and (289) $M^{\perp}$ and $\eta_n^{\perp}$, from (302) and (289) $M^{\times}$ and $\eta_n^{\times}$, and from (294) k. With these parameters determined we can calculate the course of macroscopic strain and mesoscopic stresses and strains for any given program of $\bar{\sigma}_{ij}$.

If necessary, more information can be drawn from the plastic part of the stress-strain diagram. Thus, e.g. let us assume that we have at our disposal the elastic-plastic stress-strain diagram corresponding to a uniaxial increasing loading $\bar{\sigma}_{12}$. In the elastic range it holds according to (259), (258) and (256):

$$\sigma_{12n} = M^{\times}\bar{\sigma}_{12}, \quad \sigma_{ijn} = 0 \quad \text{for} \quad ij \neq 12, 21 \tag{303}$$

$$\sigma_{12n}' = \frac{(v_e\mu_n + v_n\mu_e)M^{\times} - \mu_e}{\mu_n}\bar{\sigma}_{12} \tag{304}$$

$$\bar{\varepsilon}_{12} = \left[v_n(\mu_n - \mu_e)M^{\times} + \mu_e\right]\bar{\sigma}_{12} \tag{305}$$

where $M^{\times}$ is defined by eq. $(289)_1$.

At the plastic limit it follows from eqs. (294). (291) and (258):

$$\bar{\sigma}_{12}^{L} = k/M^{\times}, \quad \sigma_{12n}^{L} = k,$$

$$\sigma_{12n}'^{L} = v_e\eta_n^{\times}\frac{\mu_n - \mu_e}{\mu_n\eta_n^{\times} + v_e\mu_e}k \tag{306}$$

and in the elastic-plastic range we have with regard to (291):

$$\sigma_{12n}\, d\sigma_{12n} = 0, \quad d\sigma_{12n} = 0$$

and substituting for $d\sigma_{12n}$ from eq. (259) ($\dot{h} = d\lambda/dt$) we get:

$$d\lambda = \frac{1}{v_e}\cdot\frac{\mu_n\eta_n^{\times} + v_e\mu_e}{v_e\sigma_{12n} - \sigma_{12n}'}d\bar{\sigma}_{12} \tag{307}$$

With this expression used in (258) and (256) we arrive at:

$$d\sigma_{12n}' = \frac{\eta_n^{\times}}{v_e}d\bar{\sigma}_{12} \tag{308}$$

$$d\bar{\varepsilon}_{12} = \frac{(v_e \mu_e + v_n \mu_n \eta_n^x)k - v_e \mu_e \sigma'_{12n}}{v_e (v_e k - \sigma'_{12n})} d\bar{\sigma}_{12} \tag{309}$$

The above equations can easily be integrated which leads to:

$$\sigma'_{12n} = \sigma''^{L}_{12n} + \frac{\eta_n^x}{v_e}(\bar{\sigma}_{12} - \bar{\sigma}^L_{12}) = \frac{\eta_n^x}{v_e}(\bar{\sigma}_{12} - k) \tag{310}$$

$$d\bar{\varepsilon}_{12} = \mu_e \frac{\bar{\sigma}_{12} - \left(1 + v_n \frac{\mu_n}{\mu_e} + \frac{v_e}{\eta_n^x}\right)k}{\bar{\sigma}_{12} - \left(1 + \frac{v_e^2}{\eta_n^x}\right)k} d\bar{\sigma}_{12} \tag{311}$$

$$\bar{\varepsilon}_{12} - \bar{\varepsilon}^L_{12} = \mu_e \left[\bar{\sigma}_{12} - \bar{\sigma}^L_{12} - k v_n \left(\frac{\mu_n}{\mu_e} + \frac{v_e}{\eta_n^x}\right) \ln \frac{\bar{\sigma}_{12} - \left(1 + \frac{v_e^2}{\eta_n^x}\right)k}{\bar{\sigma}^L_{12} - \left(1 + \frac{v_e^2}{\eta_n^x}\right)k}\right] \tag{312}$$

From the structure of the above equations it is clear that with $\bar{\varepsilon}_{12}$ increasing $\bar{\sigma}_{12}$ approaches the value

$$\bar{\sigma}_{12}^{max} = \left(1 + \frac{v_e^2}{\eta_n^x}\right)k \tag{313}$$

and the stress-strain curve approaches the direction parallel with the $\bar{\varepsilon}_{12}$-axis. The fact that the values of $\bar{\sigma}_{12}$ are limited for $\bar{\varepsilon}_{12}$ growing over all limits is of course natural, as the matrix is perfectly elastic-plastic and the elastic material constituent forms inclusions /fibers/. For the fictitious case of homogeneous stress ($\eta_n^x = \infty$ in (313)) $\bar{\sigma}_{12}^{max} = k$ and for the other fictitious case of homogeneous strain ($\eta_n^x = 0$) $\bar{\sigma}_{12}^{max}$ is infinite.

For a numerical example we will use the experimental data published by K.M. Prewo and K.G. Kreider 61/. The described material is a composite with aluminium matrix and silicon carbide coated boron fibers /BORSIC/.

In our notation the experimental input data are:

$E_n = 10 \times 10^6$ psi
$E_e = 58 \times 10^6$ psi
$v_n = 0.48$
$\bar{E}^{\perp} = 21.3 \times 10^6$ psi

and the experimental stress-strain diagram for tension $\bar{\sigma}_{22}$ in the direction normal to the fibers, as depicted in Fig. 22 by individual points.

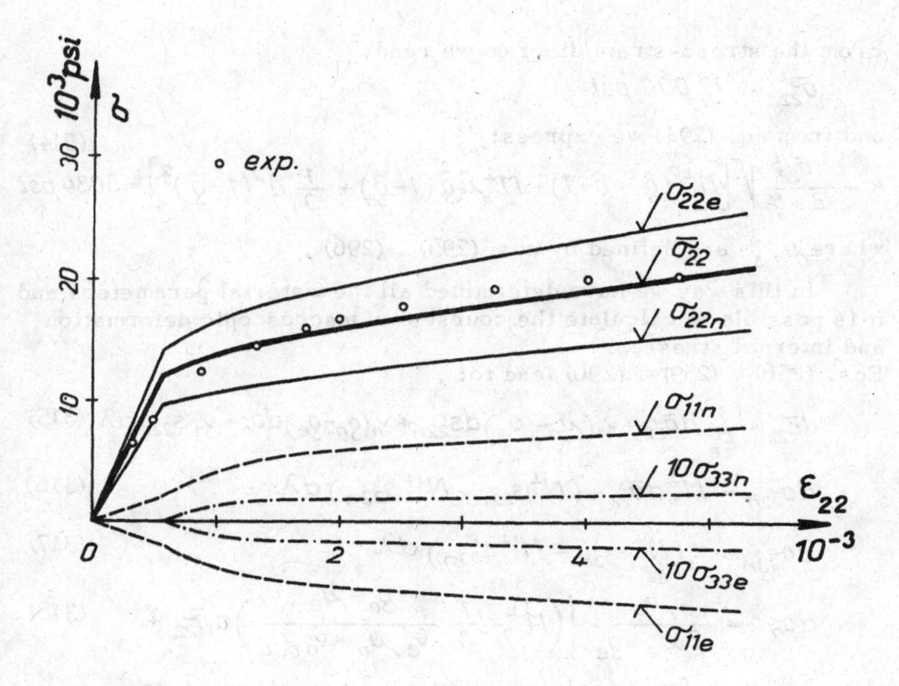

Fig. 22 Internal stresses in a fiber- reinforced material under uniaxial tension in the direction normal to the fibers; elastic fibers, elastic-plastic matrix. /Exp. points from K.M. Prewo and K.G. Kreider [61]//.

The common Poisson's ratio for both the material constituents was assumed in the value of 0.33.

By definition we have:

$$\mu_e = \frac{1+\nu}{E_e} = 2.2931 \times 10^{-8}\ psi^{-1}$$

$$\mu_n = \frac{1+\nu}{E_n} = 1.33 \times 10^{-7}\ psi^{-1}$$

$$\varrho_e = \frac{1-2\nu}{E_e} = 5.8621 \times 10^{-9}\ psi^{-1}$$

and from (299) and (289) :

$$M^{\perp} = 0.80427$$

$$\eta_n^{\perp} = 0.82985$$

From the stress-strain diagram we read:

$$\bar{\sigma}_{22}^{L} = 12\,000 \text{ psi}$$

and from eq. (294) we express:

$$k = \frac{\bar{\sigma}_{22}^{L}}{2+\hat{\varrho}} \sqrt{\left[M^{\perp 2}(\hat{\varrho}^2+\hat{\varrho}+1) + M^{\perp}\hat{\mu}\hat{\varrho}(1-\hat{\varrho}) + \frac{1}{3}\hat{\mu}^2(1-\hat{\varrho})^2\right]} = 5084 \text{ psi} \tag{314}$$

where $\hat{\mu}$, $\hat{\varrho}$ are defined by eqs. (295) , (296) .

In this way we have determined all the material parameters and it is possible to calculate the courses of macroscopic deformation and internal stresses.

Eqs. (256) , (259) , (290) lead to:

$$d\bar{\varepsilon}_{22} = \frac{1}{E_e} d\bar{\sigma}_{22} + v_n(\mu_n - \mu_e) ds_{22n} + v_n(\varrho_n - \varrho_e) d\sigma_n + v_n s_{22n} d\lambda \tag{315}$$

$$d\sigma_{22n} = M^{\perp} d\bar{\sigma}_{22} - (N^{\perp} s_{22n} - N'^{\perp} s'_{22n}) d\lambda \tag{316}$$

$$d\sigma_{33n} = -(N^{\perp} s_{33n} - N'^{\perp} s'_{33n}) d\lambda \tag{317}$$

$$d\sigma_n = \frac{\mu_e}{2\mu_e + \varrho_e} \left\{ \left(M^{\perp} + \frac{1}{3} \frac{\varrho_e - \mu_e}{v_e \mu_n + v_n \mu_e} \right) d\bar{\sigma}_{22} + \right. \tag{318}$$

$$\left. + \left[s_{11n} \left(N^{\perp} - \frac{v_e}{v_e \mu_n + v_n \mu_e} \right) - N'^{\perp} s'_{11n} \right] d\lambda \right\}$$

$$d\sigma_{11n} = -d\sigma_{22n} - d\sigma_{33n} + 3 d\sigma_n \tag{319}$$

With the expressions for $d\sigma_{11n}$, $d\sigma_{22n}$, $d\sigma_{33n}$ known we can calculate $d\bar{\varepsilon}_{22}$, $d\sigma'_{11n}$, $d\sigma'_{22n}$, $d\sigma'_{33n}$ from eqs. (315) , (258) . The only unknown that remains to be determined is $d\lambda$. This can easily be done with the use of eq. (291) , where we assume strain-hardening in the form $k = k_0 + a\lambda$.

After differentiation this equation yields for our case:

$$s_{11n} ds_{11n} + s_{22n} ds_{22n} + s_{33n} ds_{33n} = 2ka\, d\lambda =$$

$$= 2a \sqrt{s_{11n}^2 + s_{22n}^2 + s_{11n} s_{22n}}\, d\lambda$$

and with the application of expressions (316) to (319) we finally derive:

$$d\lambda\left\{\frac{s_{11n}^2}{2\mu_e+\varrho_e}\left[\frac{3v_e\mu_e}{v_e\mu_n+v_n\mu_e}+(\mu_e+2\varrho_e)N^{\perp}\right]+2N^{\perp}s_{22n}\cdot\right.$$

$$\cdot(s_{11n}+s_{22n})-N'^{\perp}\left[s'_{11n}\left(s_{11n}\frac{\mu_e+2\varrho_e}{2\mu_e+\varrho_e}+s_{22n}\right)+\right.$$

$$\left.\left.+s'_{22n}(s_{11n}+2s_{22n})\right]+2a\sqrt{s_{11n}^2+s_{22n}^2+s_{11n}s_{22n}}\right\}=$$

$$=d\bar{\sigma}_{22}\left[s_{11n}\frac{\varrho_e-\mu_e}{2\mu_e+\varrho_e}\left(\frac{\mu_e}{v_e\mu_n+v_n\mu_e}-M^{\perp}\right)+s_{22n}M^{\perp}\right] \tag{320}$$

With all these expressions available it is now easy to plot the courses of macroscopic and internal stresses using a small calculator and increasing the value of $\bar{\sigma}_{22}$ by small steps.
The resulting curves are shown in Fig.22. The values of a and $\bar{\sigma}_{22}^L$ were determined with a try and error procedure to get the best possible agreement with the experimental points of the $\bar{\sigma}_{22}-\bar{\varepsilon}_{22}$ diagram. The value of a determined in this way is:

$$a=2\times10^9\,(psi)^2$$

Concluding this paragraph let us shortly comment on the transition to the homogeneous stress model. In sections I.2.1 , I.2.2 , I.2.3 the transition is straightforward: infinite values of $\eta_{\alpha\beta n}$ or vanishing values of $\chi_{\alpha\beta n}$ for all possible combinations of the indices lead to the homogeneous stress model. In section I.2.4 /unidirectional continuous fibers/ the transition is not so simple. For the x_1 -direction /direction of the fibers/ there are no adequate structural parameters and instead of eq. (287) we have eq. (286). Therefore. complete transition to homogeneous stress model for all stress components is impossible to achieve by a variation of the structural parameters. In special cases only, where eq. (286) is not used /e.g. eq. (303) or (313) / this possibility exists.

In the elastic range the transition to a homogeneous stress field as well as homogeneous strain field can of course be arrived at if equalling the elastic compliances:

$$\mu_e = \mu_n \tag{321}$$

Thus, e.g. if using (321) in (295) and (289) we get $\hat{\mu} = M^{\perp} = 1$ and eq. (314) simplifies to:

$$k = \frac{\bar{\sigma}_{22}^{L}}{\sqrt{3}} = \frac{\sigma_{22n}}{\sqrt{3}}$$

in accordance with (291) and (259) .

I.2.5 Final Remarks on the Model for Materials with Macroscopic Transverse Isotropy

Some results concerning materials with macroscopic transverse isotropy were published earlier in author's papers 28,30,36,37/ .
Let us comment on some differences between these papers and the results presented here.

First of all there appeared unfortunately some misprints in 30/ : the formulae appearing there sub numbers (4.17) , (4.18) are not correct, the right formulae are presented here sub number (271) ; furthermore, the section 4. of the quoted paper does not relate to a reinforcement by discrete particles, as can be read in its title, but to a reinforcement by a continuous lattice.

The mathematical models of materials with unidirectional continuous fibers, as they are presented in 28/ and 30/, can be - according to our today's understanding - simplified: the unidirectional continuous fibers can be described as a special case of inclusions, i.e. the respective structural parameters η_f, η'_f appearing in the quoted papers, can be taken as infinite. All the formulae can easily be changed in this way and then they correspond to the formulae presented here.

The formulae and procedures advanced here in the preceding paragraphs have the character of demonstrative examples. Many other variants can be constructed in analogous ways. Thus, e.g. in the case of unidirectional fibers we have shown the solution to the identification problem based on the experimental elastic-plastic stress-strain curves, whereas in the case of the reinforcement by a three-dimensional lattice the solution based on the viscoelastic flow-curve. Of course that similar solutions can be worked out for the unidirectional fibers in a viscoelastic matrix and for the threedimensional lattice in an elastic-plastic matrix. The model must be adapted to the concrete material in question and to the concrete loading conditions. The method of doing so was explained in what preceded.

There remains the question of elastic Poisson's ratios. In the final formulae Poisson's ratios were assumed equal for both the material constituents. This was not because without this simplifying assumption it would not be possible to derive analogous formulae. Anybody can do it without problems. But the formulae are substantially more complicated, the solution to the identification problem substantially more difficult and the difference in the results that can be expected does not seem to be worth undergoing such trouble. Anybody who worked out concrete calculations of inelastic deformations and compared them with experiments knows that the exactness that can be achieved is not high and that it is justified to consider the effect of different elastic Poisson's ratios negligeable. On top of it the exactness with which Poisson's ratios can be measured and considered constant, is also problematic. On the contrary it seems reasonable in many cases to accept the more crude assumption of Poisson's ratios equalling 0.5, i.e. of elastic incompressibility. The respective formulae - substantially simpler - follow from those presented above in a straightforward way.

II. MESOMECHANICAL LIMIT ANALYSIS

In the first part of this monograph we tried to construct such a mathematical model, for which the solution to the identification problem would be easy and the necessary input information simple to acquire. Nevertheless it was necessary to determine the constants of the constitutive equations of the material constituents and the structural parameters. Then it was possible to calculate the course of macroscopic strain and mesoscopic stresses and strains for a given loading. Here we have in mind such an approach, where the course of stresses and strains in the deformation process cannot be determined, only some limits for the resulting macroscopic strength, but the necessary input information is substantially simpler. This approach is called mesomechanical limit analysis, as it relates the strength criteria on the mesoscale with those on the macroscale.

When speaking about the limit analysis of heterogeneous materials we have in mind the ultimate macroscopic stress or strain that the materials are able to bear. As the quantities under question are tensors, it is necessary to investigate their limits as hypersurfaces in the stress- or strain-space, similarly as it is usual in the case of yield-criteria.

Some problems of this kind - as far as the limits in the stress space are concerned - have been solved on the basis of the well-known theorems proved by Drucker, Prager and Greenberg 10/. Applications of these theorems to composite materials are outlined by Majumdar and McLaughlin 49/ which contains a number of further quotations.

In the presented paper a different approach is shown that enables the deduction of the upper bounds of strength in the stress- and strain-space under very simple assumptions: not only is it not necessary to suppose perfect plasticity, but the constitutive equations of the material constituents may be quite general and so can be the initial internal stresses and strains.

The deduction and the results are simple, they encompass an unlimited number of material constituents and make it possible to express the limits in the stress space as well as in the strain space.

The results presented here follow the reasoning presented in author's paper 32/, which is a generalization of a previous author's paper 31/.

II. 1 The Upper Bound Expressed in Terms of Stresses

Let us suppose that the microparticles of the heterogeneous material under question can be categorized into N material kinds characterized by their material properties. These kinds will be called material constituents.

The material is supposed to be in static equilibrium and thus it holds:

$$\sum_{n=1}^{N} v_n \langle \sigma_{ij} \rangle_n = \bar{\sigma}_{ij} \tag{322}$$

where v_n is the volume fraction of the n-th material constituent, $\langle \sigma_{ij} \rangle_n$ is the mean value of the stress tensor in the n-th material constituent and $\bar{\sigma}_{ij}$ is the macroscopic stress.

Then we can prove:

Theorem I: "Let the upper bound of the strength of any n-th material constituent be expressed by

$$A_{ijkl} (\sigma_{ij} - \overset{n}{\xi}_{ij})(\sigma_{kl} - \overset{n}{\xi}_{kl}) \leq c_n^2 \tag{323}$$

where the matrix A_{ijkl} is such that

$$A_{ijkl} = A_{klij} \quad \text{and} \quad A_{ijkl}\, t_{ij}\, t_{kl}$$

is positive-definite for any symmetrical t_{ij}.

Then the upper bound for the heterogeneous material in terms of macroscopic stress is

$$A_{ijkl} \left(\bar{\sigma}_{ij} - \sum_{n=1}^{N} v_n \overset{n}{\xi}_{ij}\right)\left(\bar{\sigma}_{kl} - \sum_{n=1}^{N} v_n \overset{n}{\xi}_{kl}\right) \leq \left(\sum_{n=1}^{N} v_n c_n\right)^{2} \text{"} \tag{324}$$

Proof

In the first step let us prove that

$$A_{ijkl} (\langle \sigma_{ij} \rangle_n - \overset{n}{\xi}_{ij})(\langle \sigma_{kl} \rangle_n - \overset{n}{\xi}_{kl}) \leq c_n^2 \tag{325}$$

To that aim the stress in any n-th material constituent will be decomposed into the mean value and the fluctuation

$$\sigma_{ij} = \langle \sigma_{ij} \rangle_n + \tilde{\sigma}_{ij} \tag{326}$$

with

$$\langle \tilde{\sigma}_{ij} \rangle_n = 0 \tag{327}$$

Then the inequality (323) takes on the form:

$$c_n^2 \geq A_{ijkl}(\sigma_{ij} - \Xi_{ij}^n)(\sigma_{kl} - \Xi_{kl}^n) = A_{ijkl}(\langle\sigma_{ij}\rangle_n + \tag{328}$$
$$+ \tilde{\sigma}_{ij} - \Xi_{ij}^n)(\langle\sigma_{kl}\rangle_n + \tilde{\sigma}_{kl} - \Xi_{kl}^n) = A_{ijkl}(\langle\sigma_{ij}\rangle_n - \Xi_{ij}^n)(\langle\sigma_{kl}\rangle_n - \Xi_{kl}^n) +$$
$$+ A_{ijkl}\tilde{\sigma}_{ij}\tilde{\sigma}_{kl} + A_{ijkl}[(\langle\sigma_{ij}\rangle_n - \Xi_{ij}^n)\tilde{\sigma}_{kl} + \tilde{\sigma}_{ij}(\langle\sigma_{kl}\rangle_n - \Xi_{kl}^n)]$$

Inequality (328) is valid in any point of the n-th material constuent and therefore it must be valid also in the average:

$$c_n^2 \geq A_{ijkl}(\langle\sigma_{ij}\rangle_n - \Xi_{ij}^n)(\langle\sigma_{kl}\rangle_n - \Xi_{kl}^n) + \langle A_{ijkl}\tilde{\sigma}_{ij}\tilde{\sigma}_{kl}\rangle_n \tag{329}$$

Summation in all the expressions is to be performed on the indexes $i,j,k,l\ (= 1, 2, 3)$ in the case of their repetition; on other indexes only if it is expressed by the symbol Σ .

With the properties of the matrix A_{ijkl} defined above is the expression $A_{ijkl}\tilde{\sigma}_{ij}\tilde{\sigma}_{kl}$ positive-definite and the inequality (325) thus follows from (329).

Furthermore the expression

$$A_{ijkl}(\langle\sigma_{ij}\rangle_n - \Xi_{ij}^n)(\langle\sigma_{kl}\rangle_p - \Xi_{kl}^p)$$

will be taken for a scalar product in a Hilbert space

$$(\boldsymbol{U}_n, \boldsymbol{U}_p)$$

with

$$\boldsymbol{U}_n = \{\langle\sigma_{ij}\rangle_n - \Xi_{ij}^n\}$$

It is an easy task to prove that all the necessary postulates of a Hilbert space are fulfilled:

$$(\boldsymbol{U}_n, \boldsymbol{U}_p) = (\boldsymbol{U}_p, \boldsymbol{U}_n)$$
$$(\boldsymbol{U}_n, \boldsymbol{U}_p + \boldsymbol{U}_q) = (\boldsymbol{U}_n, \boldsymbol{U}_p) + (\boldsymbol{U}_n, \boldsymbol{U}_q)$$
$$(a\boldsymbol{U}_n, \boldsymbol{U}_p) = a(\boldsymbol{U}_n, \boldsymbol{U}_p)$$
$$(\boldsymbol{U}_n, \boldsymbol{U}_n) \geq 0$$
$$(\boldsymbol{U}_n, \boldsymbol{U}_n) = 0 \text{ if and only if } \boldsymbol{U}_n = \boldsymbol{0}$$

Then the respective form of Schwarz's inequality says that

$$A_{ijkl}(\langle\sigma_{ij}\rangle_n - \Xi_{ij}^n)(\langle\sigma_{kl}\rangle_p - \Xi_{kl}^p) \leq \tag{330}$$
$$\leq \sqrt{[A_{ijkl}(\langle\sigma_{ij}\rangle_n - \Xi_{ij}^n)(\langle\sigma_{kl}\rangle_n - \Xi_{kl}^n)]}\sqrt{[A_{ijkl}(\langle\sigma_{ij}\rangle_p - \Xi_{ij}^p)(\langle\sigma_{kl}\rangle_p - \Xi_{kl}^p)]}$$

From (325) and (330) we conclude:

$$A_{ijkl}(\langle\sigma_{ij}\rangle_n - \Xi_{ij}^n)(\langle\sigma_{kl}\rangle_p - \Xi_{kl}^p) \leq c_n c_p \tag{331}$$

and with regard to (322) and (331) we have finally

$$A_{ijkl}(\bar{\sigma}_{ij} - \sum_{n=1}^{N} v_n \Xi_{ij}^{n})(\bar{\sigma}_{kl} - \sum_{n=1}^{N} v_n \Xi_{kl}^{n}) = \tag{332}$$

$$= A_{ijkl}\left[\sum_{n=1}^{N} v_n(\langle\sigma_{ij}\rangle_n - \Xi_{ij}^{n})\right]\left[\sum_{n=1}^{N} v_n(\langle\sigma_{kl}\rangle_n - \Xi_{kl}^{n})\right] =$$

$$= \sum_{n,p=1}^{n,p=N} v_n v_p A_{ijkl}(\langle\sigma_{ij}\rangle_n - \Xi_{ij}^{n})(\langle\sigma_{kl}\rangle_p - \Xi_{kl}^{p}) \leqq$$

$$\leqq \sum_{n,p=1}^{n,p=N} v_n v_p c_n c_p = \left(\sum_{n=1}^{N} v_n c_n\right)^2$$

which proves Theorem I.

The parameters Ξ_{ij}^{n} make it possible to formulate different values of strength in tension and in compression. For e.g. $A_{1111} = 1$, $\Xi_{11}^{n} = c_n$ the limits for uniaxial stress in the x_1-direction will be

$$(\sigma_{11} - c_n)^2 \leqq c_n^2$$

or

$$0 \leqq \sigma_{11} \leqq 2c_n$$

i.e. zero strength in compression and finite strength in tension, which is a property that can be attributed to textile fibers. On the contrary, some other materials may display high strength in compression and low strength in tension, which is again easy to describe by the parameters Ξ_{ij}^{n} with opposite signs.

If the fundamental inequality (323) is expressed in deviatoric components, the proof may be performed in the same way as for the normal components replacing in all the expressions σ_{ij} by s_{ij}. This follows from the fact that equation (322) is valid also for deviatoric parts of the stress tensor, which is very easy to prove: making traces of the tensors in equation (322) and substracting the resulting equation divided by three from the original equation we obtain

$$\sum_{n=1}^{N} v_n \langle s_{ij}\rangle_n = \bar{s}_{ij}$$

with

$$s_{ij} = \sigma_{ij} - \delta_{ij}\frac{1}{3}\sigma_{kk}$$

where δ_{ij} is Kronecker's delta.

An important special case is the Mises' criterion which will be arrived at if inequality (323) is written in terms of the deviators and if we introduce

$$A_{ijkl} = \delta_{ik}\delta_{jl}, \quad \Xi_{ij}^{n} = 0$$

From Theorem I it then follows that for the strength of the material constituents being limited by Mises' criteria

$$s_{ij} s_{ij} \leq c_n^2 \tag{333}$$

the strength of the heterogeneous material is limited by

$$\bar{s}_{ij} \bar{s}_{ij} \leq (v_1 c_1 + v_2 c_2 + \ldots + v_N c_N)^2 \tag{334}$$

The last inequality was published for the first time without proof by Z. Hashin [14] with the note that it was deduced by the author in an unpublished paper on the basis of the classical limit theorems /i.e. for the case of perfectly elastic-plastic materials/. Under broader assumptions, proceeding similarly as in this paper, the author has deduced it in [31].

However, it is well known that if the Mises criterion is used as a criterion of strength, it must be completed by another criterion, as the first invariant of the stress tensor cannot increase without limits. The simplest complementary criterion can be arrived at if introducing

$$A_{ijkl} = \delta_{ij} \delta_{kl} \tag{335}$$

in (323). This gives:

$$(\sigma_{ii} - \xi_{ii}^n)(\sigma_{kk} - \xi_{kk}^n) \leq c_n^2 \tag{336}$$

or simpler

$$(\sigma - \hat{\xi}_n)^2 \leq \hat{c}_n^2$$

where

$$\sigma = \frac{1}{3} \sigma_{ii} \; .$$

This means:

$$-\hat{c}_n + \hat{\xi}_n \leq \sigma \leq \hat{c}_n + \hat{\xi}_n$$

The respective form of (324) then reads:

$$\left(\bar{\sigma} - \sum_{n=1}^{N} v_n \hat{\xi}_n\right)^2 \leq \left(\sum_{n=1}^{N} v_n \hat{c}_n\right)^2 \tag{337}$$

or

$$-\sum_{n=1}^{N} v_n \hat{c}_n + \sum_{n=1}^{N} v_n \hat{\xi}_n \leq \bar{\sigma} \leq \sum_{n=1}^{N} v_n \hat{c}_n + \sum_{n=1}^{N} v_n \hat{\xi}_n \tag{338}$$

From experimental evidence we know that $\hat{\xi}_n < 0$, $\hat{c}_n > 0$.

II.2 The Upper Bound Expressed in Terms of Strains

In some cases it may be preferable to express the limits in terms of strains or to express them in terms of stresses with a subsidiary condition that the deformation cannot overpass certain limits.

In the case of continuous displacements it follows from the compatibility condition that

$$\sum_{n=1}^{N} v_n \langle \varepsilon_{ij} \rangle_n = \bar{\varepsilon}_{ij} \tag{339}$$

i.e. an equation in terms of strains which is quite analogous to equation (322) for stresses.

The proof of the Theorem I, valid for stresses, may therefore be performed step by step for strains and we arrive at a quite similar theorem:

Theorem II: "Let the upper bound of the strength of any n-th material constituent be expressed by

$$B_{ijkl}(\varepsilon_{ij} - \psi_{ij}^{n})(\varepsilon_{kl} - \psi_{ij}^{n}) \leq C_n^2 \tag{340}$$

where the matrix B_{ijkl} is such that

$$B_{ijkl} = B_{klij}$$

and

$$B_{ijkl}\, t_{ij}\, t_{kl}$$

is positive-definite for any symmetrical t_{ij}.

Then the upper bound for the heterogeneous material in terms of the macroscopic strains is

$$B_{ijkl}\left(\bar{\varepsilon}_{ij} - \sum_{n=1}^{N} v_n \psi_{ij}^{n}\right)\left(\bar{\varepsilon}_{kl} - \sum_{n=1}^{N} v_n \psi_{ij}^{n}\right) \leq \left(\sum_{n=1}^{N} v_n C_n\right)^{2}\text{"} \tag{341}$$

The matrix B_{ijkl} is again the same for all material constituents and their specific properties are characterized by the parameters C_n and ψ_{ij}^{n} only.

Similarly again as in the case of stresses is it possible to formulate such a theorem for deviatoric parts of strains.

II.3 The Lower Bounds

In the case of material constituents being perfectly elastic-plastic with unlimited ductility the well-known theorems of Drucker, Prager and Greenberg [10] may be applied, homogeneous stress may be chosen for the statically admissible field:

$$\sigma_{ij} = \bar{\sigma}_{ij}$$

and with the real limits of the strength of the material constituents given by (323) the macroscopic lower bound will be represented by the inner envelope of the hypersurfaces

$$A_{ijkl}(\bar{\sigma}_{ij} - \xi^n_{ij})(\bar{\sigma}_{kl} - \xi^n_{kl}) = c^2_n \qquad (n = 1, 2, \dots, N) \tag{342}$$

If the limits of all the constituents are expressed by the Mises' criteria (333) the macroscopic lower bound will be

$$\bar{s}_{ij}\,\bar{s}_{ij} = c^2_{min} \tag{343}$$

where c_{min} is the minumum $c\ (n = 1, 2, \dots, N)$ among these parameters of the material constituents /cf. Hashin 14//.

In the general case, where plasticity is not assumed, the formulation of the lower bound is a very difficult task. To clarify this problem a bit let us consider a two-phase material with the constituents being supposed brittle, their strength being given by Mises' criteria (333) and the structure formed by a weaker matrix and stronger inclusions.

Let a representative volume of this composite be loaded on its surface uniformly by :

$$p_i = \bar{\sigma}_{ij}\, n_j \tag{344}$$

where $\bar{\sigma}_{ij}$ is the macroscopic stress, constant in space, and n_j is the exterior unit normal.

Let us further suppose the unfavourable situation that there exists a plane surface S_c that cuts only the weaker material - the matrix- and divides the representative volume into two parts. Let the outer surface of one of these parts be denoted by S_1. Then the condition of static equilibrium of this part gives:

$$\int_{S_1} \bar{\sigma}_{ij}\, n_j\, dS + \int_{S_c} \sigma_{ij}\, n_j\, dS = 0 \tag{345}$$

where n_j are the exterior unit normals to the separated part on the outer surface S_1 and on the cutting plane surface S_c.

It is clear enough that $\bar{\sigma}_{ij}$ on S_1 and n_j on S_c are constant in space and therefore we may rewrite equation (345) as follows:

$$\bar{\sigma}_{ij}\int_{S_1} n_j\, dS + n_j \int_{S_c} \sigma_{ij}\, dS = \bar{\sigma}_{ij}\int_{S_1} n_j\, dS + \langle\sigma_{ij}\rangle_{Sc}\int_{S_c} n_j\, dS = 0 \tag{346}$$

where $\langle \sigma_{ij} \rangle_{Sc}$ is the mean value of stress on S_c.

The surface $(S_1 + S_c)$ is closed and it holds therefore:

$$\int_{S_1} n_j \, dS + \int_{S_c} n_j \, dS = 0 \tag{347}$$

and thus

$$\langle \sigma_{ij} \rangle_{Sc} = \bar{\sigma}_{ij} \tag{348}$$

Forming trace in (348) and subtracting its third from the original equation we get

$$\langle s_{ij} \rangle_{Sc} = \bar{s}_{ij} \tag{349}$$

Let us further assume that everywhere on S_c the limit of strength is reached, i.e. it holds in any point on S_c:

$$s_{ij} \, s_{ij} = c_m^2 \tag{350}$$

with c_m meaning the parameter corresponding to the matrix.

We will decompose further the deviatoric stress on S_c into the mean value and the fluctuation:

$$s_{ij} = \langle s_{ij} \rangle_{Sc} + \tilde{s}_{ij} \tag{351}$$

From (350) and (351) it follows

$$c_m^2 = \langle s_{ij} \rangle_{Sc} \langle s_{ij} \rangle_{Sc} + \tilde{s}_{ij} \tilde{s}_{ij} + 2 \langle s_{ij} \rangle_{Sc} \tilde{s}_{ij} \tag{352}$$

and averaging over S_c:

$$c_m^2 = \langle s_{ij} \rangle_{Sc} \langle s_{ij} \rangle_{Sc} + \langle \tilde{s}_{ij} \tilde{s}_{ij} \rangle_{Sc} \tag{353}$$

From equations (353) and (349) we can finally conclude

$$\bar{s}_{ij} \, \bar{s}_{ij} < c_m^2 \tag{354}$$

for s_{ij} not constant on S_c.

Contrary to the materials with plastic deformation, in this case even such macroscopic stress may be dangerous which lies inside the inner envelope of the hypersurfaces of the local strength limits of the material constituents.

It is clear from what has been shown how dangerous may be the combination of only brittle constituents and of what great significance plasticity is for the strength of composite materials. It is an apparent paradox that between two otherwise identical composite materials. that one may be stronger whose one or more constituents have lower strenght in terms of stress but they are able to deform plastically. Consequently the strenght of the same composite may be higher under higher temperature.

III. A NON-LOCAL CRITERION OF STRENGTH

M. Hlaváček and V. Kafka

III. 1 Foreword

The problem of mathematical modelling of strength is far more complicated than the mathematical modelling of elastic or inelastic deformation. There are two fundamental approaches to the subject: the continuum mechanics approach and the fracture mechanics approach.

In the simplest variant of the former the fields of stresses or strains are calculated and the points are found where their values are most dangerous. If the characters of the material and of the loading process are such that before rupture internal changes appear. the application of the continuum damage mechanics or of the microrheology. describing changes of internal microstresses. can be useful. Nevertheless there are important phenomena that can hardly be described in this way: as two simple examples let us mention a/ the difference between the tensile strength in pure tension and that in bending, b/ the differing relative strength of fibers with differing diameters.

Fracture mechanics is able to get closer to reality with its prediction of strength, taking into account the existing cracks and the stress field round them, but unfortunately the situation, form, extent and orientation of the cracks are usually not known and the defectoscopic measurement can be afforded only exceptionally.

This led us looking for an alternative approach that would adequately explain and model the phenomena that are out of reach of the classical continuum mechanics, an approach that would work with input information that is easy to acquire. The factors that influence the strength are very complicated and numerous. For an operative mathematical model we must select only the most important of them. Our approach is based on the classical continuum mechanics calculations of the stress field but two points more are included that are considered the most important:

a/ the influence of the surface layer
b/ the non-local influence of the surrounding stress field in some neighbourhood of the fracturing locus.

The concrete form of our mathematical model and its possibilities of describing a number of observed phenomena will be advanced later on. Before coming to it. let us comment on some other attempts to explain the two above mentioned cases /the differing strength in pure tension and in bending, and the differing relative strength of fibers with differing diameters/.

One of the explanations that can be found for both these phenomena is the probability of appearance of a weak spot. The influence of this factor exists without any doubt. But it is far from being as strong as the differences observed. Thus, e.g. Puch [62/] reports experimental results according to which a 332 times longer glass fiber displayed only 2 times lower strength. On the other hand if the length remained unchanged and the diameter of the fiber was 7,6 times larger, its strength was 9 times lower. It is clear enough that there is no correlation either in volumes or in surfaces - as areas of the possible weak spots.

In the case of axisymmetrical bending of plates of anorganic glass the factor of the weak spot is given by the state of the surface under tension, where fracture always begins, not by the volume. This is proved beyond any doubt by experiments with scratched surfaces. Hence. in plates with different thicknesses but with the same quality of the surfaces and the same dimensions, the probability of appearance of a weak spots is the same. But in spite of it the observed strength differs with the thickness, is higher at thin plates - in accordance with the consequences of our concept.

As to the strength of the fibers a number of other explanations were advanced too:

- The influence of the surface layer. But the surface layer is generally weaker than the inner material and the volume fraction of the surface layer is higher at thinner fibers, which would make thinner fibers weaker.
- The influence of internal stresses. But internal stresses cannot influence the strength of materials that can deform plastically - small plastic deformation is sufficient (cf. e.g. Drucker. Prager, Greenberg [10/]). And the influence of the diameter is observed even with such materials.
- The influence of orientation of the crystalline lattice. But the effect under consideration is observed even in whiskers, where the orientation is unique (cf. Bowden, Tabor [5/]).
- The influence of higher strain - hardening in thinner fibers. But higher strain - hardening affects the elastic limit, not the strength.
- The influence of microcracks that are assumingly smaller in thinner fibers and larger in thicker fibers. This explanation is difficult to contradict if it is true. But as far as our knowledge reaches there exist neither sufficient evidence nor serious reasons for it. In any case thinner fibers are more deformed during their production which could cause larger microcracks; the process of production and the material properties are very different in different materials but the observed effect is similar; the effect is observed even on the scale of diameters

in 10^{-6}m (cf. 5/) where it is difficult to imagine the existence of the different cracks.

Hence, the very different explanations that have been offered up to now do not seem to the authors satisfactory. In what follows we advance another - energetic - explanation of this and of many other phenomena that leads to a simple quantitative mathematical model.

III.2 Criterion For Isotropic Materials

The explanation and description of the above mentioned phenomena can be found in the fact that the creation of a macrocrack does not depend on the local quantities alone, but also on the stress - field in the neighbourhood of the point in question. The ways, in which this influence may be described, can be different. We believe that the advantage of the way we have chosen is its generality and simplicity:

With the exception of an explosion, fracture is always anisotropic in its essential nature. Therefore, the driving part of the active stress tensor is its deviatoric part rather than its isotropic part. The latter can be seen as a scalar parameter that influences only the scalar features of the fracturing process, its effect differs from the effect of the deviatoric part similarly as in plasticity, viscosity etc.

As an illustration of this idea let us compare brittle fracture under uniaxial tension $\sigma_{11}>0$ and brittle fracture under biaxial compression $\sigma_{22}=\sigma_{33}<0$. In the first case the deviatoric and the isotropic parts are:

$$s_{ij}=\frac{\sigma_{11}}{3}\begin{pmatrix}2&0&0\\0&-1&0\\0&0&-1\end{pmatrix},\quad \sigma_{ij}=\frac{\sigma_{11}}{3}\begin{pmatrix}1&0&0\\0&1&0\\0&0&1\end{pmatrix}$$

In the second case:

$$s_{ij}=-\frac{\sigma_{22}}{3}\begin{pmatrix}2&0&0\\0&-1&0\\0&0&-1\end{pmatrix},\quad \sigma_{ij}=\frac{2\sigma_{22}}{3}\begin{pmatrix}1&0&0\\0&1&0\\0&0&1\end{pmatrix}.$$

It is clear that the deviatoric parts are similar and for $\sigma_{11}=-\sigma_{22}$ identical, and really - the planes of fracture are also similar - normal to the x_1-axis. On the other hand the isotropic parts differ substantially - one negative, the other positive, and therefore the scalar measure of the deviatoric stress (e.g. $\sqrt{s_{ij}\ s_{ij}}$) in the fracturing critical state will be different: $|\sigma_{22}^c|>|\sigma_{11}^c|$.

Very often the plane of a crack is assumed to be normal to the highest principal stress component of the acting stress tensor, and the creation of the crack is supposed to be in causal connection with this stress component. However, this scheme can lead to a paradox. e.g. in the case of the biaxial compression mentioned above, where the highest principal stress component is zero, and it is difficult to imagine that a zero component is in causal connection with the crack that really appears. We have no such trouble if we accept the scheme that the driving factor is not the acting tensor itself, but its deviatoric part. The highest principal stress component of the deviator is always positive, i.e. tension, and if we assume that the plane of crack is normal to this component, we get the same direction as in the first scheme /normality to the highest principal stress component of the whole tensor/, but there is no paradox.

The simplest possibility to take into account the two influences - that of the deviatoric part and that of the isotropic part - is a linear combination of two scalars with physical meaning - of the respective elastic energies:

$$(W_d)_P + k' \langle W_i \rangle_\varrho \leq C \qquad (355)$$

where $(W_d)_P$ is the elastic energy density of the deviatoric part of stress at the fracturing locus P, W_i the elastic energy density of the isotropic part of stress; $\langle --- \rangle_\varrho$ means the average value in the spherical neighbourhood with radius ϱ; k', C, are constants that differ for the inner points from those for the surface points /where they depend on the state of the surface/.

Furthermore, two cases with different k', C must be distinguished:

$$(\sigma_{ii})_P > 0 \quad \text{and} \quad (\sigma_{ii})_P < 0 \, .$$

For these two cases C is different, but always positive, $k' > 0$ in the former and $k' < 0$ in the latter case. This corresponds to the notion that the isotropic extension or compression in the ϱ -neighbourhood of the point in question reduces or increases the strength, respectively.

If the point in question is the tip of a crack, C is supposed to be smaller if the crack started moving.

If a part of the ϱ -neighbourhood is a free space outside the body, the area of the ϱ -neighbourhood is not influenced, but the density of elastic energy in this free space part is of course vanishing and thus it lowers the mean value of the density in the ϱ -neighbourhood.

If it exceptionally happens that $(\sigma_{ii})_P > 0$ $\left((\sigma_{ii})_P < 0 \right)$, but in a part of the ϱ -neighbourhood $\sigma_{ii} < 0 \; (\sigma_{ii} > 0)$, this part is to be considered as free of energy.

Very often the points of maximum exposure are not isolated points, but a set of points forming a line or a surface. In such cases the

ϱ -neighbourhood, which is a sphere for an isolated point, is to be considered a cylinder with radius ϱ for a line and a layer of thickness 2ϱ for the case of a surface of points with maximum exposure.

It is one of the main points of the approach that the density of the deviatoric energy in eq. (355) is represented by its value in the fracturing locus P, whereas the density of the isotropic part by the mean value in some ϱ -neighbourhood. We have no proof for such a formula. We have only some indications and agreement with experimental results:

In the elastic state the difference between the stress-strain relations for the deviatoric parts and the isotropic parts is only quantitative, but for the inelastic processes - plastic, viscous or fracturing /all of them mean violation of bonds in the material on some scale/ - the difference is also qualitative. This difference substantiates also the assumption that the effective macroscopic values - which are defined as averages of microscopic values over some area - are to be evaluated over different areas for the deviatoric and the isotropic parts.

To illustrate this point of view let us model at first a very simple axisymmetrical case - a circle that is a part of an unlimited plane. A deviatoric axisymmetrical violation of bonds in the circle can be demonstrated by rotation of concentric annuli that are parts of the circle. It is clear that such a process has a local character: it need not influence the stress-field in the plane. Another situation arises if the axisymmetric violation of bonds in the circle has an isotropic character, i.e. an isotropic rise of distances among the particles of the circle. Then the stress-field in the neighbourhood of the circle will necessarily be influenced and vice versa - the isotropic stress field in the neighbourhood will influence the violation of bonds in the circle.

It is also well known that a plastic deformation process - that is deviatoric - is described by differential equations of a hyperbolic type allowing local processes, whereas elastic deformation - that includes isotropic parts - is described by differential equations of an elliptical type where any local change is felt in the whole body. /But of course the influence of the first invariant is negligeable in classical plasticity - contrary to fracturing processes/.

These considerations indicate that in criterion (355) the influence of the deviatoric energy can be represented by the local value of its density in the respective point only, whereas the influence of the isotropic part by a mean value in some larger neighbourhood.

For large specimens /the order of their dimensions higher than the order of ϱ / under homogeneous stress the value of ϱ becomes irrelevant and the deviatoric and isotropic parts of the stress tensor can be written as follow s:

$$s_{ij} = s\, S_{ij} \quad , \quad \sigma_{ij} = \sigma\, \delta_{ij}$$

where δ_{ij} is the Kronecker delta and

$$\delta_{ij}\, S_{ij} = 0 \quad , \quad S_{ij}\, S_{ij} = 1$$

For a surface point P on a flat surface criterion (355) can now be written:

$$(W_d)_P + k' \langle W_i \rangle_\varrho = (W_d)_P + \frac{1}{2} k' (W_i)_P = as^2 + b\sigma^2 \leq C$$

and according to the assumptions about k', C it is

$$a > 0\ ,\ C > 0\ , \qquad \begin{array}{ll} b > 0 & \text{for } \sigma > 0 \\ b < 0 & \text{for } \sigma < 0 \end{array}$$

The graphical representation of the above strength criterion is shown in Fig. 23.

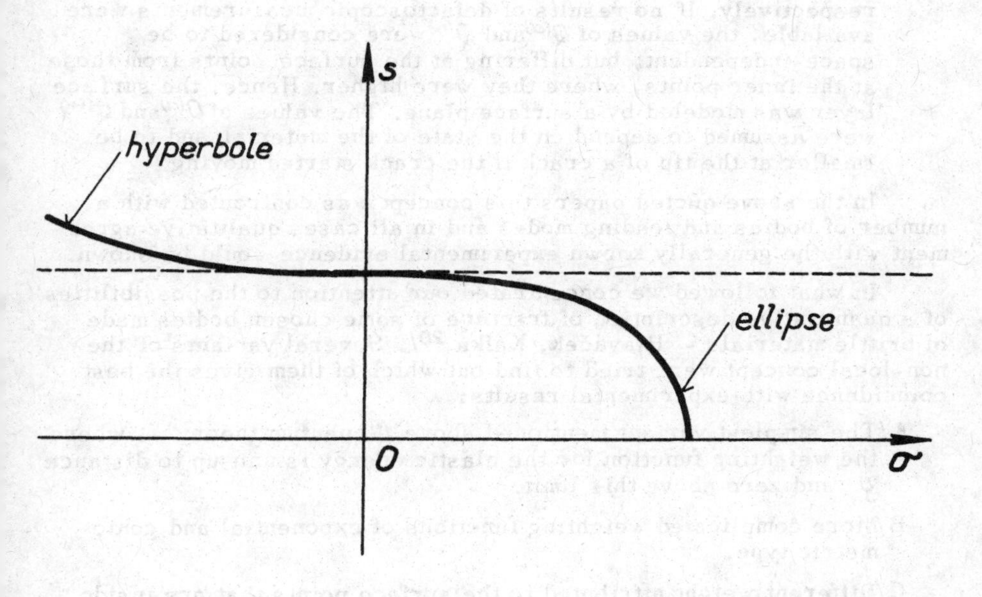

Fig. 23 Graphical representation of the strength criterion (355) if applied to large specimens under homogeneous stress.

This can be accepted as a simple approximative formula for the description of the well known experimental evidence. However, the advantages of the criterion turn out in cases, where the stress state is not homogeneous and/or where ϱ is not negligeable in regard to the dimensions of the body.

Before concluding this paragraph let us mention that the way to this last form (355) of the criterion was not straightforward.

In preceding communications Kafka [39,41]/Kafka, Němec [40]/ a model called "quantum theory of strength" was advanced that can be looked upon as the simplest variant of a non-local theory:

a/For isotropic materials the effective neighbourhood was specified as the set of points with their distance from the fracturing locus not overpassing a certain limit ϱ. The limit ϱ was considered a constant that depends on the material and its structure.

b/The effect of the stress-field in the ϱ-neighbourhood was specified by two scalar functions - the deviatoric part $W_e^{(d)}$ of the elastic energy and the tensile isotropic part $W_e^{(i+)}$ of the elastic energy - that could not overpass certain limits $Q^{(d)}$, $Q^{(i+)}$ respectively. If no results of defectoscopic measurements were available, the values of $Q^{(d)}$ and $Q^{(i+)}$ were considered to be space-independent, but differing at the surface points from those at the inner points, where they were higher. Hence, the surface layer was modeled by a surface plane. The values of $Q^{(d)}$ and $Q^{(i+)}$ were assumed to depend on the state of the material and to be smaller at the tip of a crack if the crack started moving.

In the above quoted papers this concept was confronted with a number of bodies and loading modes and in all cases qualitative agreement with the generally known experimental evidence could be shown.

In what followed we concentrated our attention to the possibilities of a quantitative description of fracture of some chosen bodies made of brittle materials - Hlaváček, Kafka [20]/. Several variants of the non-local concept were tried to find out which of them gives the best coincidence with experimental results:

A/The simplest variant mentioned above /"quantum theory"/, where the weighting function for the elastic energy is unit up to distance ϱ and zero above this limit.

B/More complicated weighting functions of exponential and goniometric type.

C/Different weight attributed to the surface points that are inside the ϱ-neighbourhood and different to the inner points of the ϱ-neighbourhood.

D/A linear combination of the "quantum theory" and the local criterion separately for the deviatoric and the isotropic parts:

$$W_P^{(i+)} + k^{(i+)} \langle W^{(i+)} \rangle_\varrho \leq C^{(i+)}$$

and

$$W_P^{(d)} + k^{(d)} \langle W^{(d)} \rangle_\varrho \leq C^{(d)}$$

where $W^{(i+)}/W^{(d)}(W_P^{(i+)}/W_P^{(d)})$ meant the density of the tensile isotropic/deviatoric part of the elastic energy (at the fracturing point P), $\langle \text{---} \rangle_\varrho$ meant the average value in the ϱ-neighbourhood and $k^{(i+)}, k^{(d)}, C^{(i+)}, C^{(d)}$ were constants.

E/ The final criterion advanced here - a linear combination of the non-local term for isotropic parts and the local criterion for deviatoric parts /eq. (355)/.

In the time we prepared the preceding papers only the criteria A, B, C, D were estimated and for the studied materials. bodies and static loading modes best results were obtained for the criterion mentioned sub D. Surprisingly, more complicated weighting functions did not lead to any appreciable improvement.

Later on we found criterion E /eq. (355)/ that can describe the cases studied in Hlaváček. Kafka [20]/as well as criterion D. but it has some serious advantages on top of it:

a/ It is simpler: only one formula instead of two.

b/ It agrees with the fact that the non-local effect does not appear in cases where the first invariant of the stress tensor is vanishing /shear of rivets x tension of wires/.

c/ For large specimens /ϱ negligeable/ the resulting formula /Fig. 23/ agrees better with experimental evidence than the Mises cylinder with a plane top. which results from D.

III. 3 Qualitative Evidence

If we compare the consequences of the non-local criterion with different generally known experimental results in a qualitative way only. the fundamental feature is the existence of a fixed length constant ϱ that is not connected with the shape and size of the body, but with the material and its structure. Having this in mind we can easily demonstrate qualitative agreement in a number of bodies and loading modes:

a/ The difference between the strength in pure tension and in tension at pure bending. - It is clear enough that with the surface tension equal in both cases the elastic energy concentrated in the ϱ-neighbourhood of the surface points is higher in the case of pure tension. Hence the higher strength in the case of bending. According to the model as well as to experimental evidence this effect is lowered with increasing plastic deformation, as at bending the stress field is homogenized at the surface.

b/ The creation of cracks in the matrix of a composite at the contact with large inclusions. - The maximum stress at the contact with one inclusion in an unlimited medium does not depend on the dimensions of the inclusion; however, the quantum of energy in the ϱ-neighbourhood of a contact point will be higher at a larger inclusion, as the dropping of stress with the distance from the contact will be slower. This explains also the lower strength of composites with coarser particles.

c/ The higher relative tensile strength of very thin bodies. - If at least one of the dimensions of the body is smaller than ϱ, then the energy comprised in any ϱ-neighbourhood is smaller, as a part of the ϱ-neighbourhood contains free space without energy. Consequently the relative strength, measured by stress, is higher. The special case of this phenomenon is the well known diameter - dependent tensile strength of fibers. This explains and describes also the high strength of fiber - reinforced composites.

d/ Such a phenomenon is not observed if wires are cut /e.g. shear strength of rivets/. This agrees with our criterion (355) where a vanishing first invariant of the stress tensor means that the non-local effect is absent.

e/ The higher relative strength in uniaxial compression of thicker samples /cf. Pisarenko. Amelianovicz, Goralik, Brezneva, Moskalenko 59//. Here the non-local effect of the elastic energy of the first invariant of the stress tensor is similar to the case of tensile strength, but not so high and in the opposite sense.

f/ The discontinuous stepwise progress of a crack. - If the energy comprised in the ϱ-neighbourhood of the tip of a crack reaches its critical value in an elastic state, then the crack starts propagating with increasing velocity, as the critical energy that is necessary for the continuation is lower than the critical energy for the onset /brittle fracture/. If the energy in the ϱ-neighbourhood reaches its critical value in a plastic state, then it drops with the progress of the crack towards the unplastified zone and after some distance the motion stops. After some time-interval a new plastification and homogenization of the stress field raises the energy in the ϱ-neighbourhood and a new step begins.

g/ Slowing-down of the progress of a crack at the surface of a body. - The ϱ-neighbourhood of the points at the tip of a crack that are near to the surface of the body comprises partly free space outside the body and therefore the energy comprised is low. In the case that the progress of the crack is parallel with the surface, this leads to the arching of the crack front; in the case that the crack approaches the surface in a perpendicular direction, this leads to the slowing down of the progress near the surface.

III.4 Quantitative Evidence

In the figures that follow the critical stresses $\tilde{\sigma}_c$ calculated according to our concept for properly chosen ϱ, k', C are compared with the mean values of $\tilde{\sigma}_c$ found experimentally.

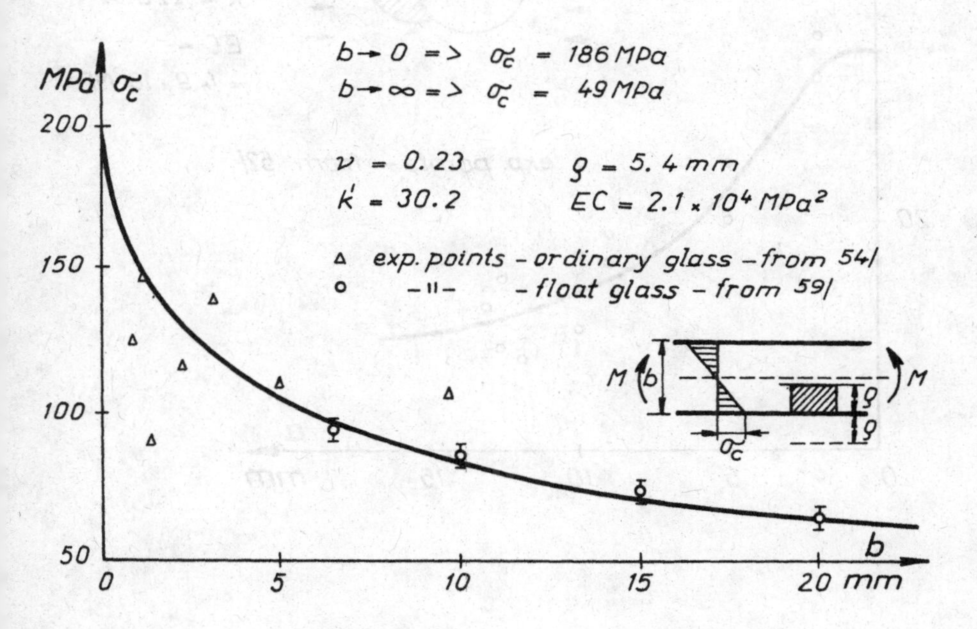

Fig. 24 Axisymmetric bending of plates - inorganic glass

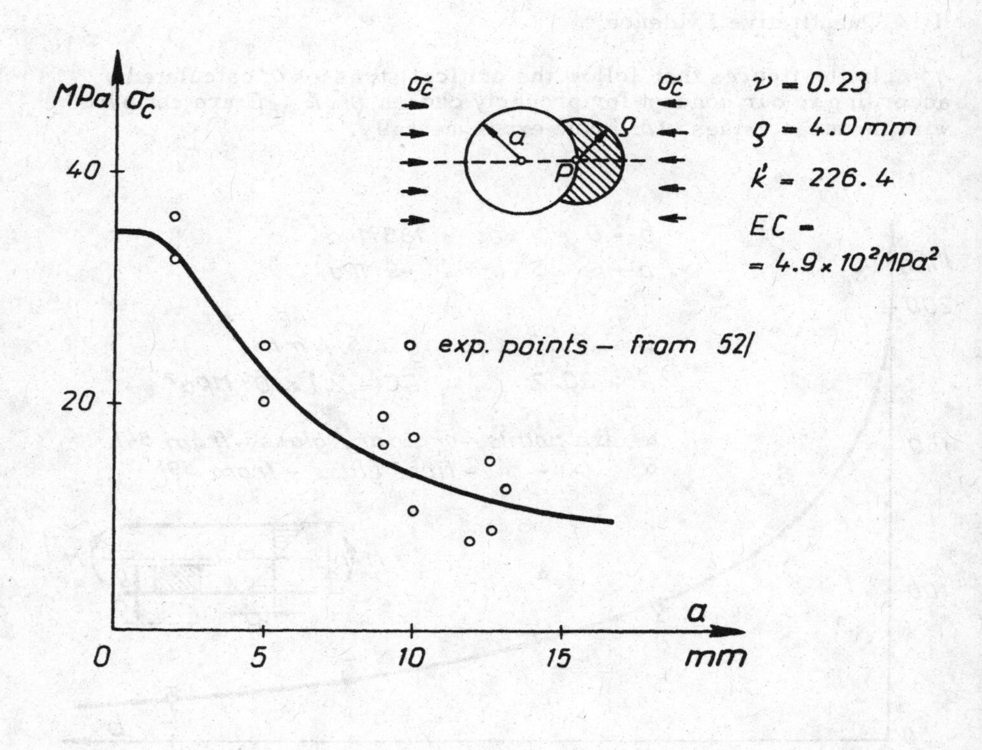

Fig. 25 Uniaxial compression of a square plate with a circular hole - inorganic glass

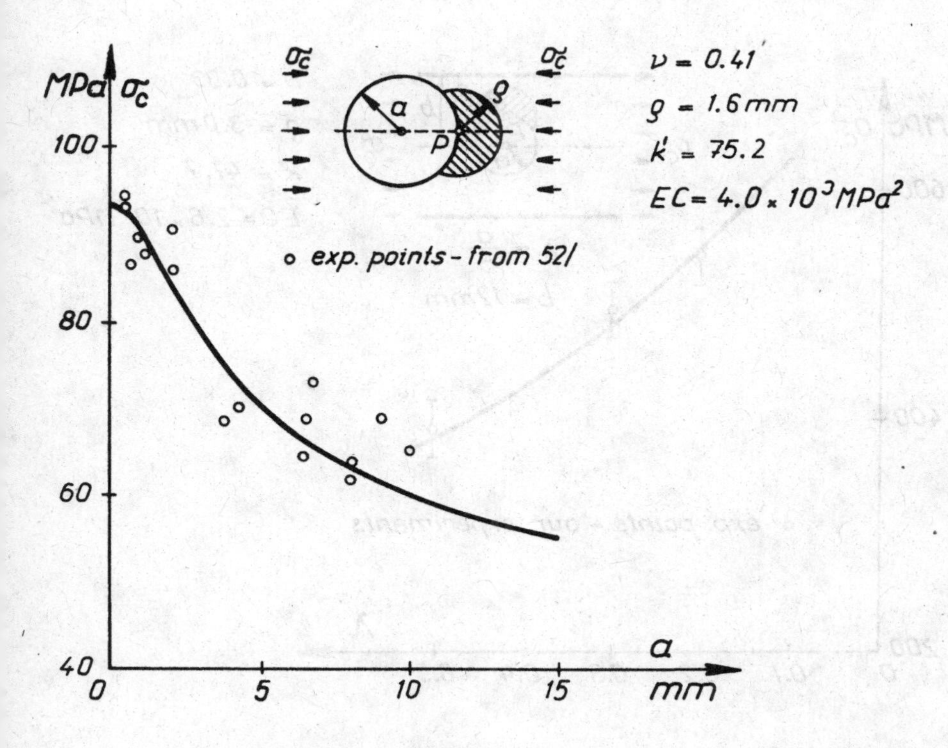

Fig. 26 Uniaxial compression of a square plate with a circular hole - organic glass /PMMA/

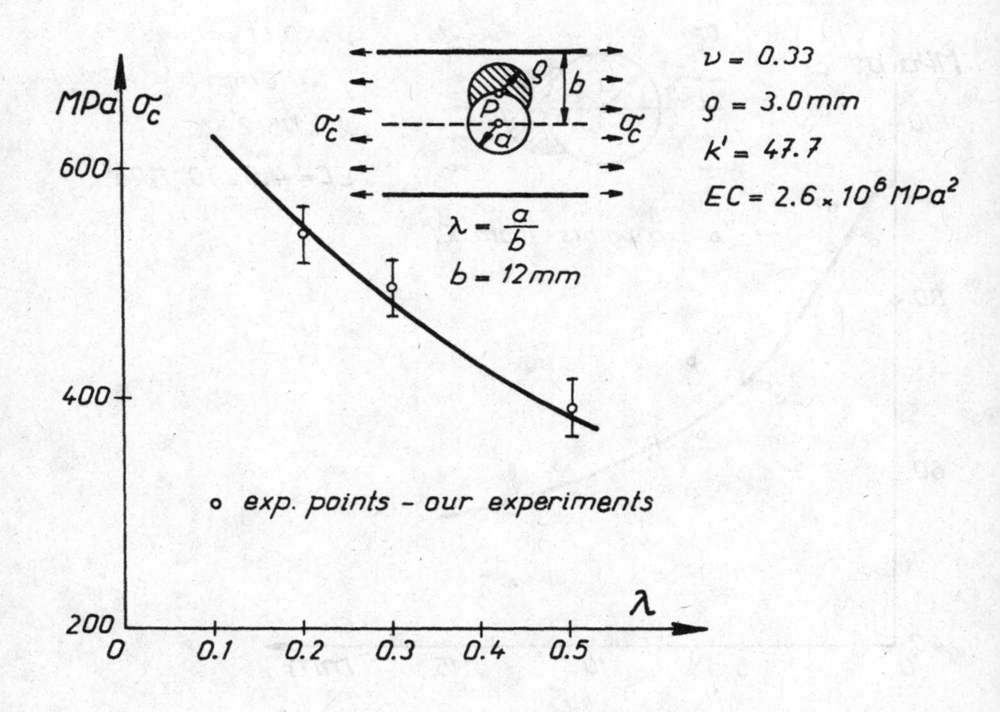

Fig. 27 Uniaxial tension of a strip with a circular hole - high strength steel " Stabil"

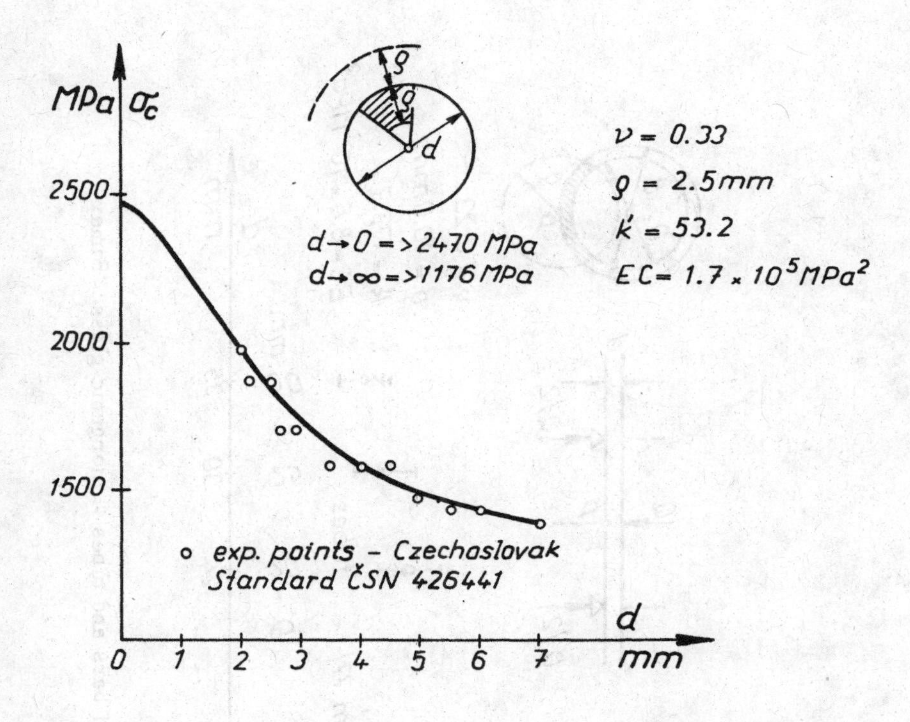

Fig. 28 Tension of wires - steel drawn wires for prestressing reinforcement

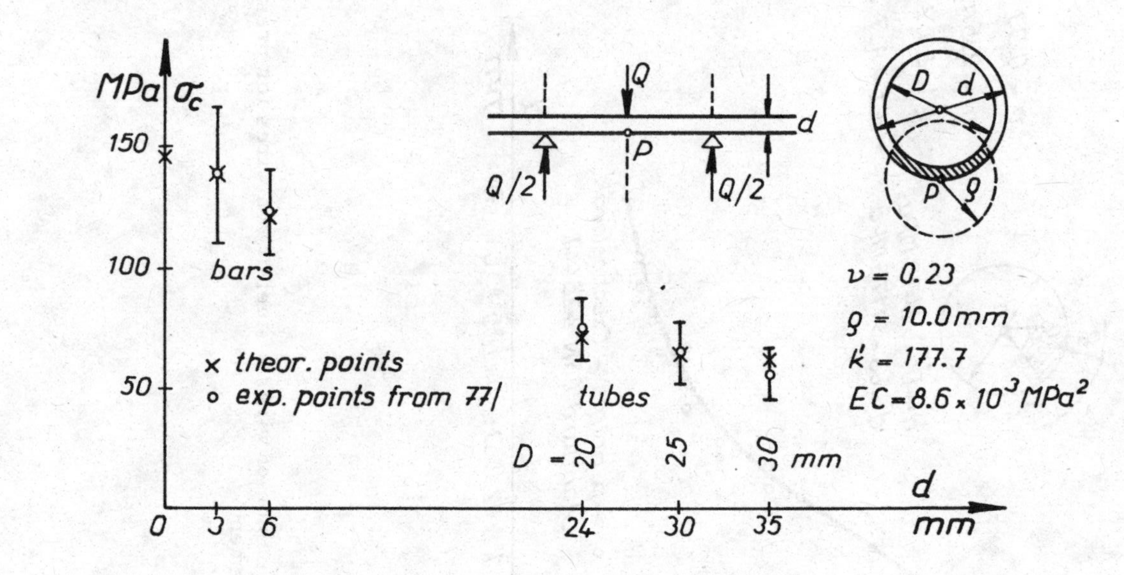

Fig.29 Three-point bending of bars and tubes - inorganic glass "Simax"

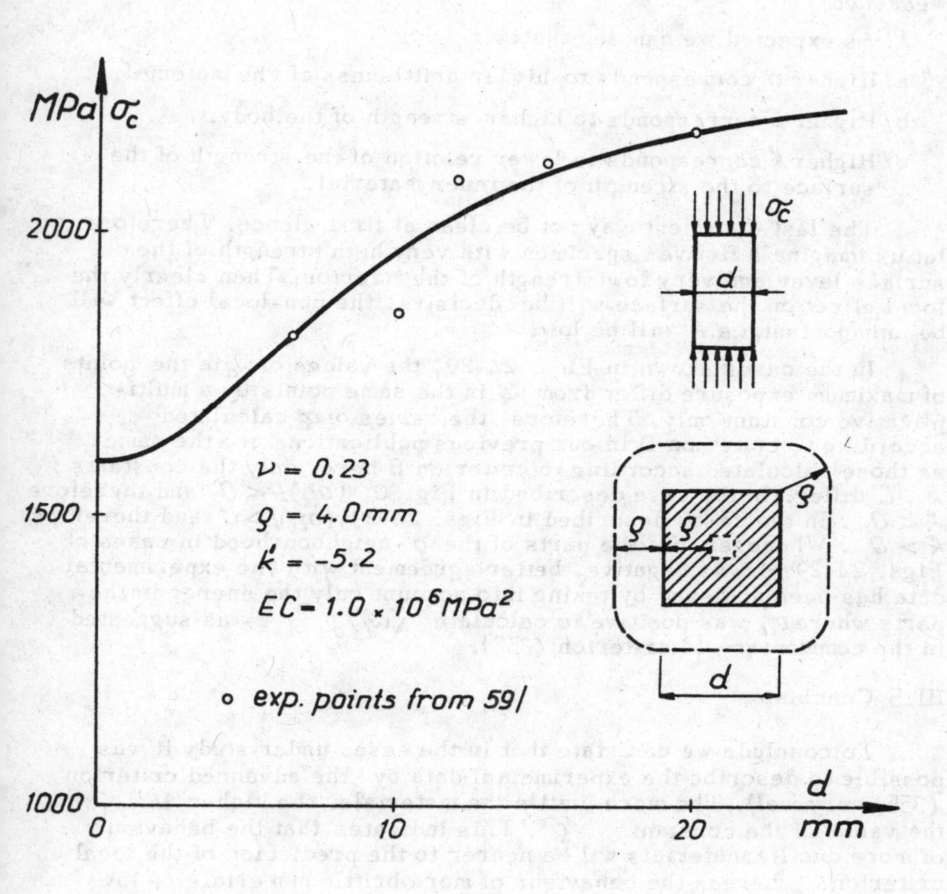

Fig. 30 Uniaxial compression of prismatic specimens - inorganic glass

Different types of surfaces explain the high differences in k' and C at similar materials /anorganic glass/. If there exists an edge, as e.g. in the case described in Fig. 25, the surface is substantially weakened.

As expected we can see that:

a/ Higher ϱ corresponds to higher brittleness of the material.

b/ Higher C corresponds to higher strength of the body.

c/ Higher k' corresponds to lower relation of the strength of the surface to the strength of the inner material.

The last statement may not be clear at first glance. Therefore let us imagine a fictive specimen with very high strength of the surface layer and very low strength of the interior. Then clearly the local effect on the surface will be decisive, the non-local effect will be unimportant, i.e. k' will be low.

In the cases shown in Figs. 24-29, the values of w_d in the points of maximum exposure differ from w_i in the same points by a multiplicative constant only. Therefore, the values of σ_c' calculated according to criterion D in our previous publications are the same as those calculated according to criterion E here, only the constants k', C differ. In the case described in Fig. 30, $(\sigma_{ii}')_P < 0$ and therefore $k' < 0$. In the cases described in Figs. 24-29, $(\sigma_{ii}')_P > 0$ and therefore $k' > 0$. Whenever in some parts of the ϱ-neighbourhood in cases of Figs. 24-29 σ_{ii}' was negative, better agreement with the experimental data has been achieved by taking into account only the energy in the parts where σ_{ii}' was positive to calculate $\langle w_i \rangle_\varrho$, - as suggested in the commentary to criterion (355).

III.5 Conclusion

To conclude we can state that in the cases under study it was possible to describe the experimental data by the advanced criterion (355) very well. The more brittle the materials, the higher is the value of the constant ϱ. This indicates that the behaviour of more ductile materials wil be nearer to the prediction of the local criterion, whereas the behaviour of more brittle materials /+ low temperature, + high rates of loading/ nearer to that of the "quantum theory of strength". How far general is the applicability of the concept, is an open question for future work.

This approach is rather new and further work may bring new development. However, it seems that on its level of simplicity formula (355) gives the best agreement with experimental evidence.

It is interesting to note that a number of biological materials /bones, bamboo and many others/ are optimized by evolution from the point of view of our concept: the surface layer is strong, but its neighbourhood has low modulus so that the energy concentrated in the neighbourhood is moderate.

Appendix 1

Proof of the statement:

„If $\langle [f(\mathbf{x})_n]^2 \rangle_n = 1$ and $\langle f(\mathbf{x})_n \rangle_n = 1$,

then $f(\mathbf{x})_n \equiv 1$."

Let $f(\mathbf{x})_n = 1 + F(\mathbf{x})_n$.

$$\langle f(\mathbf{x})_n \rangle_n = \frac{1}{V_n} \int_{V_n} [1 + F(\mathbf{x})_n] dV = 1 + \frac{1}{V_n} \int_{V_n} F(\mathbf{x})_n \, dV = 1$$

and therefore $\int_{V_n} F(\mathbf{x})_n \, dV = 0$.

$$\langle [f(\mathbf{x})_n]^2 \rangle_n = \frac{1}{V_n} \int_{V_n} [1 + F(\mathbf{x})_n]^2 dV = \frac{1}{V_n} \int_{V_n} [1 + 2F(\mathbf{x})_n +$$

$$+ F(\mathbf{x})_n^2] \, dV = 1 + \frac{1}{V_n} \int_{V_n} F^2(\mathbf{x})_n \, dV = 1$$

and therefore $F(\mathbf{x})_n \equiv 0$ and $f(\mathbf{x})_n \equiv 1$

q.e.d.

Appendix 2

Deduction of eqs. (46) and (47).

The first step will be deduction of eq. $(46)_2$ from eqs. (38) to (45). Let us note that eqs. (38), (39), (40) can be rewritten in quite similar forms separately for deviatoric and isotropic parts. Furthermore, eqs. (38). (39), (40), (41), (42) and (43) are valid throughout the deformation process and therefore they are valid in the same form also for the rates of stress, strain and T. Having this in mind we can easily proceed as follows:

$$\dot{\bar{\varepsilon}} = v_e \dot{\varepsilon}_e + v_n \dot{\varepsilon}_n$$

$$\dot{\varepsilon}_e = \varrho_e \dot{\sigma}_e + \alpha_e \dot{T}$$

$$\dot{\varepsilon}_n = \varrho_n \dot{\sigma}_n + \dot{\varrho}_n \sigma_n + \dot{\tau} + \alpha_n \dot{T}$$

$$\dot{\sigma}_e = \frac{1}{v_e} (\dot{\bar{\sigma}} - v_n \dot{\sigma}_n)$$

$$\dot{\bar{\varepsilon}} = v_e \left[\varrho_e \frac{1}{v_e} (\dot{\bar{\sigma}} - v_n \dot{\sigma}_n) + \alpha_e \dot{T} \right] +$$

$$+ v_n (\varrho_n \dot{\sigma}_n + \dot{\varrho}_n \sigma_n + \dot{\tau} + \alpha_n \dot{T}) =$$

$$= \dot{\bar{\sigma}} \varrho_e + \dot{\sigma}_n v_n (\varrho_n - \varrho_e) + \sigma_n v_n \dot{\varrho}_n +$$

$$+ v_n \dot{\tau} + \dot{T} (v_e \alpha_e + v_n \alpha_n)$$

$$\dot{\sigma}_n = \dot{\sigma}_e + \frac{\dot{\sigma}_e'}{\eta_e^o} - \frac{\dot{\sigma}_n'}{\eta_n^o} =$$

$$= \frac{1}{v_e} (\dot{\bar{\sigma}} - v_n \dot{\sigma}_n) + \frac{\dot{\sigma}_e'}{\eta_e^o} - \frac{\dot{\sigma}_n'}{\eta_n^o}$$

from which:

$$\dot{\sigma}_n = \dot{\bar{\sigma}} + v_e \left(\frac{\dot{\sigma}_e'}{\eta_e^o} - \frac{\dot{\sigma}_n'}{\eta_n^o} \right)$$

$$\dot{\sigma}_n' = \frac{1}{\varrho_n}\left(\dot{\varepsilon}_n' - \dot{\varrho}_n \sigma_n'\right) =$$

$$= \frac{1}{\varrho_n}\left(\dot{\varepsilon}_n - \dot{\bar{\varepsilon}} - \dot{\varrho}_n \sigma_n'\right) =$$

$$= \frac{1}{\varrho_n}\left(\dot{\varepsilon}_n - v_n \dot{\varepsilon}_n - v_e \dot{\varepsilon}_e - \dot{\varrho}_n \sigma_n'\right) =$$

$$= \frac{1}{\varrho_n}\left[v_e\left(\dot{\varepsilon}_n - \dot{\varepsilon}_e\right) - \dot{\varrho}_n \sigma_n'\right]$$

$$\dot{\sigma}_e' = \frac{1}{\varrho_e}\dot{\varepsilon}_e' = \frac{1}{\varrho_e}\left(\dot{\varepsilon}_e - \dot{\bar{\varepsilon}}\right) =$$

$$= \frac{1}{\varrho_e}\left(\dot{\varepsilon}_e - v_e \dot{\varepsilon}_e - v_n \dot{\varepsilon}_n\right) =$$

$$= \frac{v_n}{\varrho_e}\left(\dot{\varepsilon}_e - \dot{\varepsilon}_n\right)$$

$$\dot{\sigma}_n = \dot{\bar{\sigma}} + v_e\left[\frac{1}{\eta_e^0}\cdot\frac{v_n}{\varrho_e}\left(\dot{\varepsilon}_e - \dot{\varepsilon}_n\right) - \frac{1}{\eta_n^0}\cdot\frac{1}{\varrho_n}v_e\cdot\right.$$

$$\left.\cdot\left(\dot{\varepsilon}_n - \dot{\varepsilon}_e\right) + \frac{1}{\eta_n^0}\cdot\frac{\dot{\varrho}_n}{\varrho_n}\sigma_n'\right] =$$

$$= \dot{\bar{\sigma}} + \frac{v_e \sigma_n'}{\eta_n^0 \varrho_n}\dot{\varrho}_n + v_e\left(\frac{v_n}{\varrho_e \eta_e^0} + \frac{v_e}{\varrho_n \eta_n^0}\right)\left(\dot{\varepsilon}_e - \dot{\varepsilon}_n\right)$$

$$\dot{\varepsilon}_e - \dot{\varepsilon}_n = \varrho_e \dot{\sigma}_e + \alpha_e \dot{T} - \varrho_n \dot{\sigma}_n - \dot{\varrho}_n \sigma_n - \dot{\tau} - \alpha_n \dot{T} =$$

$$= \varrho_e \frac{1}{v_e}\left(\dot{\bar{\sigma}} - v_n \dot{\sigma}_n\right) - \varrho_n \dot{\sigma}_n + \left(\alpha_e - \alpha_n\right)\dot{T} - \sigma_n \dot{\varrho}_n - \dot{\tau} =$$

$$= \frac{\varrho_e}{v_e}\dot{\bar{\sigma}} - \left(\frac{v_n \varrho_e}{v_e} + \varrho_n\right)\dot{\sigma}_n + \left(\alpha_e - \alpha_n\right)\dot{T} - \sigma_n \dot{\varrho}_n - \dot{\tau}$$

$$\dot{\sigma}_n = \dot{\bar{\sigma}} + \frac{v_e\,\sigma_n'}{\eta_n^o\,\varrho_n}\dot{\varrho}_n + v_e\left(\frac{v_n}{\eta_e^o\,\varrho_e} + \frac{v_e}{\eta_n^o\,\varrho_n}\right)\cdot$$

$$\cdot\left[\frac{\varrho_e}{v_e}\dot{\bar{\sigma}} - \frac{1}{v_e}(v_e\,\varrho_n + v_n\,\varrho_e)\dot{\sigma}_n + (\alpha_e - \alpha_n)\dot{T} - \right.$$

$$\left. - \sigma_n\,\dot{\varrho}_n - \dot{\tau}\right]$$

and by a simple rearrangement of the last equation we arrive at formula $(47)_3$, which is used in the following expression of $\dot{\bar{\varepsilon}}$:

$$\dot{\bar{\varepsilon}} = \varrho_e\,\dot{\bar{\sigma}} + \sigma_n\,v_n\,\dot{\varrho}_n + v_n\,\dot{\tau} + (v_e\,\alpha_e + v_n\,\alpha_n)\,\dot{T} +$$

$$+ v_n(\varrho_n - \varrho_e)\left[M_o\dot{\bar{\sigma}} - (N_o\sigma_n - N_o'\sigma_n')\dot{\varrho}_n - N_o(\alpha_n - \alpha_e)\dot{T} - N_o\dot{\tau}\right]$$

and in a straightforward way the last formula can be rearranged to $(46)_2$.

The deduction of formula $(47)_4$ from the last expression for $\dot{\sigma}_n'$ given above is then clear without comments.

There is no need to repeat the same procedure for the deviatoric parts, as it is quite similar. Instead of $\sigma, \varepsilon, \varrho, \dot{\varrho}_n, \alpha, \dot{\tau}, M_o, M_o'$, N_o, N_o', R_o, η^o it is necessary to write $s_{ij}, e_{ij}, \mu, \hbar, 0, 0, M, M'$, N, N', R, η and we obtain formulae $(46)_1$, $(47)_{1,2}$, (48),(49).

Appendix 3

Deduction of eqs. (58) and (59).

The procedure is similar to that in Appendix 2. Again we will start with the expressions for the isotropic parts - here with eqs. $(58)_2$ and $(59)_2$.

$$\dot{\bar{\varepsilon}} = v_e \dot{\varepsilon}_e + v_n \dot{\varepsilon}_n$$

$$\dot{\varepsilon}_e = \rho_e \dot{\sigma}_e + \alpha_e \dot{T} = \frac{\rho_e}{v_e}(\dot{\bar{\sigma}} - v_n \dot{\sigma}_n) + \alpha_e \dot{T}$$

$$\dot{\varepsilon}_n = \rho_n \dot{\sigma}_n + \dot{\rho}_n \sigma_n + \dot{\tau} + \alpha_n \dot{T}$$

$$\dot{\sigma}_n = \dot{\bar{\sigma}} + \dot{\sigma}_n' = \dot{\bar{\sigma}} + \frac{1}{\rho_n}(\dot{\varepsilon}_n' - \dot{\rho}_n \sigma_n') =$$

$$= \dot{\bar{\sigma}} + \frac{1}{\rho_n}\left[\chi_n^o\left(\dot{\varepsilon}_e - \dot{\varepsilon}_n + \frac{\dot{\varepsilon}_e'}{\chi_e^o}\right) - \dot{\rho}_n \sigma_n'\right]$$

$$\dot{\varepsilon}_e - \dot{\varepsilon}_n = \frac{\rho_e}{v_e}\dot{\bar{\sigma}} - \frac{v_e \rho_n + v_n \rho_e}{v_e}\dot{\sigma}_n - \sigma_n \dot{\rho}_n +$$

$$+ (\alpha_e - \alpha_n)\dot{T} - \dot{\tau}$$

$$\dot{\varepsilon}_e' = \rho_e \dot{\sigma}_e' = \rho_e(\dot{\sigma}_e - \dot{\bar{\sigma}}) = \rho_e\left[\frac{1}{v_e}(\dot{\bar{\sigma}} - v_n \dot{\sigma}_n) - \dot{\bar{\sigma}}\right] =$$

$$= \rho_e \frac{v_n}{v_e}(\dot{\bar{\sigma}} - \dot{\sigma}_n)$$

$$\dot{\sigma}_n = \dot{\bar{\sigma}} + \frac{1}{\rho_n}\left\{\chi_n^o\left[\frac{\rho_e}{v_e}\dot{\bar{\sigma}} - \frac{v_e \rho_n + v_n \rho_e}{v_e}\dot{\sigma}_n - \sigma_n \dot{\rho}_n +\right.\right.$$

$$\left.\left. + (\alpha_e - \alpha_n)\dot{T} - \dot{\tau} + \rho_e \frac{v_n}{v_e \chi_e^o}(\dot{\bar{\sigma}} - \dot{\sigma}_n)\right] - \dot{\rho}_n \sigma_n'\right\}$$

From the last equation we express $\dot{\sigma}_n$ and arrive at $(59)_2$ with the use of eq. $(56)_2$.

$$\dot{\bar{\varepsilon}} = v_e\left[\frac{\varrho_e}{v_e}\left(\dot{\bar{\sigma}} - v_n\dot{\sigma}_n\right) + \alpha_e\dot{T}\right] + v_n\left(\varrho_n\dot{\sigma}_n + \dot{\varrho}_n\sigma_n + \alpha_n\dot{T} + \dot{\tau}\right)$$

Expression $(59)_2$ is further applied to the last equation and in this way we easily deduce by mere rearrangement eq. $(58)_2$.

Similarly as in the case of the A-model /Appendix 2/ it is not necessary to repeat the procedure for the deviatoric parts, it is sufficient to replace σ, ε, ϱ, $\dot{\varrho}_n$, α, $\dot{\tau}$, P_o, Q_o, R_o, η^o by s_{ij}, e_{ij}, μ, $\dot{\kappa}$, 0, 0, P, Q, R, η and we obtain formulae $(58)_1$ and $(59)_1$.

Appendix 4

Determination of the input data for the solution to the identification problem based on the elastic-plastic stress-strain diagram.

At first let us discuss the case of homogeneous elastic constants $(\mu_e = \mu_n = \bar{\mu}\,,\ \varrho_e = \varrho_n = \bar{\varrho})$.

The experimental basis for determining the constants $k = \sqrt{\frac{3}{4}}c$, $\nu_n = 1 - \nu_e$, η_e, η_n is the stress-strain diagram of a specimen in simple tension or in simple shear.

The determination of the constant k or c is straightforward, k being the plastic limit in simple shear $(\bar{\sigma}_{12})_L$, c the deviatoric plastic limit in simple tension $\left(c = (\bar{s}_{11})_L = \frac{2}{3}(\bar{\sigma}_{11})_L\right)$. However, the term plastic limit has a specific meaning in our concept. The mathematical model relates to a developed elastic-plastic deformation process that starts from the elastic part of the diagram by an abrupt change of the direction that is followed by a smooth curve with decreasing curvature. This may not be adequate to describe exactly the experimental curve at the beginning of the plastic part, where a yield point jog or a smooth change of the first derivative can appear. In such cases the plastic limit is to be determined as the stress corresponding to the intersection of the straight line of the elastic portion of the diagram with a backward extrapolation of the characteristic curve corresponding to the already developed plastic deformation /see point L in Fig. 13/.

To determine ν_n, η_n, η_e we must know the values of X, Y, Y' in three points of a stress-strain diagram in simple tension /eqs. (136) to (145) /. Let us measure in such three points their coordinates $\bar{\varepsilon}_{11}$, $\bar{\sigma}_{11}$ and the respective derivatives $d\bar{\sigma}_{11}/d\bar{\varepsilon}_{11}$.

Then according to eqs. (136) and (139) :

$$X = \frac{\bar{e} - (\bar{e})_L}{c\,\mu_e} = \frac{\bar{e}_{11} - (\bar{e}_{11})_L}{c\,\mu_e}$$

$$\bar{e}_{11} = \bar{\varepsilon}_{11} - \bar{\varepsilon} = \bar{\varepsilon}_{11} - \frac{1}{3}\bar{\varrho}\,\bar{\sigma}_{11}$$

$$Y = \frac{\bar{s} - (\bar{s})_L}{c} = \frac{\bar{s}_{11} - (\bar{s}_{11})_L}{c}$$

$$\bar{s}_{11} = \frac{2}{3}\bar{\sigma}_{11}$$

$$\gamma' = \mu_e \frac{d\bar{s}_{11}}{d\bar{e}_{11}} = \frac{2\mu_e}{3(d\bar{\sigma}_{11}/d\bar{\varepsilon}_{11})^{-1} - \bar{g}}$$

If the known experimental stress-strain diagram corresponds to simple shear, the above relations are quite analogous, but:

$$\bar{e} = \bar{e}_{12} = \bar{\varepsilon}_{12}$$

$$\bar{s} = \bar{s}_{12} = \bar{\sigma}_{12}$$

$$\gamma' = \mu_e \frac{d\bar{\sigma}_{12}}{d\bar{\varepsilon}_{12}}$$

and instead of c it is necessary to use k. Equations (135) to (145) are valid also with k instead of c.

In the case that the elastic constants of the two material constituents are different, the procedure complicates. All the above equations are valid. the values of $\mu_e (\neq \mu_n \neq \bar{\mu})$ and of $\bar{g}$ are supposed to be known, but it is difficult to determine c or k. We must use an iterative procedure in this case. In the first step we proceed as if the material was elastically homogeneous and determine c, v_n, η_e, η_n. Then - using the values of v_n, η_e, η_n so determined - new value of c is calculated from eq. $(47)_1$:

$$c = \frac{\mu_e \mu_n \eta_e \eta_n + \mu_e (v_e \mu_e \eta_e + v_n \mu_n \eta_n)}{\mu_e \mu_n \eta_e \eta_n + (v_e \mu_n + v_n \mu_e)(v_e \mu_e \eta_e + v_n \mu_n \eta_n)} (\bar{s}_{11})_L$$

This new value of c is used for a new calculation of v_n, η_e, η_n and the procedure goes on until the differences are small.

Appendix 5

Discussion of eqs. (219) and (220).

a/ It is easy to show that for a monotonic uniaxial loading /i.e. for a loading $\bar{s}_{ij} = \bar{s}\, T_{ij}$, where T_{ij} is constant and $\bar{s}$ is non-decreasing or non-increasing from a zero original state/ the inelastic part of $\dot{s}_e (s_{ije} = s_e T_{ij})$ has the same sign as $\dot{s}_e$ and as the elastic part of $\dot{s}_e$:

Let us start with $\dot{s}_n$. From $(67)_1$ and $(69)_{1,2}$ we get:

$$(\dot{s}_n)_{el} = (P+\bar{P})\dot{\bar{s}} = \frac{v_e \mu_e + \mu_n \eta_n}{\mu_n \eta_n + v_e (v_e \mu_n + v_n \mu_e)}\, \dot{\bar{s}}$$

The coefficient of $\dot{\bar{s}}$ is positive and therefore, the sign of $(\dot{s}_n)_{el}$ is the same as that of $\dot{\bar{s}}$. From (67) and (218) it is easy to derive the expression for $(\dot{s}_e)_{el}$ and show that it is also of the same sign as $\dot{\bar{s}}$.

It is possible to imagine that any inelastic process is composed of small steps and every step beginns with the immediate elastic response that is followed by the inelastic process that proceeds in some time-interval /which is really so/. The first elastic response leads to the following value of $(\bar{Q}\bar{s} - Q s_n)$ in eq. (67) :

$$\bar{Q}\bar{s} - Q(s_n)_{el} = -v_e^2 \mu_e \frac{v_e + \eta_n}{R^2}\, \bar{s}$$

The sign of $\dot{h}$ is always positive and therefore, the sign of the inelastic part of $\dot{s}_n$, i.e. of $(\dot{s}_n)_{nl}$, is opposite to that of $\bar{s}$ and s_n at the first elastic response. This means that the absolute value of s_n diminishes due to the inelastic part of deformation.

If the value of $\bar{s}$ is held fixed, the process of diminishing s_n can proceed, but s_n cannot overpass the value

$$s_n = \frac{\bar{Q}}{Q}\, \bar{s}$$

at which the coefficient of $\dot{h}$ in $(67)_1$ becomes zero.

If the next rise of $\bar{s}$ steps in, the elastic response leads to the rise of the absolute values of s_n and of $(\bar{Q}\bar{s} - Q s_n)$, but the sign of the elastic change of s_n coincides with that of $\bar{s}$, whereas the sign of the elastic change of $(\bar{Q}\bar{s} - Q s_n)$ is opposite. Hence, the sign of the resulting inelastic change of s_n, i.e. of $(\dot{s}_n)_{nl}$ will be again opposite to that of $\bar{s}$ and the absolute value of s_n will be diminished by the inelastic response throughout the process.

The above considerations lead to the conclusion that the signs of s_n and of $(\dot{s}_n)_{el}$ will coincide with that of $\bar{s}$, whereas the signs of $(\bar{Q}\bar{s} - Q s_n)$ and of $(\dot{s}_n)_{nl}$ will be opposite to that of $\bar{s}$. Furthermore, if $\bar{s}$ is held fixed and the process is viscous /not limited by some yield condition or some criterion of fracturing/ s_n will reach after an infinite time the value

$$(s_n)_\infty = \frac{\bar{Q}}{Q}\,\bar{s} = \frac{\eta_n}{v_e^2 + \eta_n}\,\bar{s}$$

Now it is straightforward to draw conclusions for s_e.
From eqs. (218) and (67) we see:

$$(\dot{s}_e)_{nl} = -\frac{v_n}{v_e}\,(\dot{s}_n)_{nl}$$

$$(s_e)_\infty = \frac{1}{v_e}\left[\,\bar{s} - v_n\,(s_n)_\infty\right] = \frac{v_e + \eta_n}{v_e^2 + \eta_n}\,\bar{s}$$

Hence, the signs of s_e, $(\dot{s}_e)_{el}$ and $(\dot{s}_e)_{nl}$ coincide with that of $\bar{s}$ and the last equation gives the value of s_e in a viscous process at fixed $\bar{s}$ after an infinite time.

b/ Now we will turn our attention to the isotropic parts of the stress- and strain-tensors. Without the addends containing $\dot{\tau}$ the situation would be quite analogous. But experimental evidence says that either the response of the isotropic parts is only elastic or there are both the inelastic addends - that containing $\dot{\varrho}_n$ as well as that containing $\dot{\tau}$. Moreover, the influence of that containing $\dot{\tau}$ is stronger, which means that whatever the kind of loading, the change of volume will be positive in the case of quasihomogeneous stable fracturing.
Similarly eq. $(67)_2$ gives:

$$(\dot{\sigma}_n)_{nl} = (\bar{Q}_o\,\bar{\sigma} - Q_o\,\sigma_n)\,\dot{\varrho}_n - (Q_o - \bar{Q}_o)\,\dot{\tau}$$

and as by definition

$$Q_o - \bar{Q}_o > 0$$

$(\dot{\sigma}_n)_{nl}$ is negative for $\dot{\tau}$ high enough and

$$(\dot{\sigma}_e)_{nl} = -\frac{v_n}{v_e}\,(\dot{\sigma}_n)_{nl}$$

is positive.

Appendix 6

Structural parameters of transversely isotropic materials.

If we suppose that our model is to be descriptive of a transversely isotropic material with the axis of symmetry X_1, then all directions perpendicular to X_1 must be equivalent. Let us label such an arbitrary direction by X_2^* and suppose that it deviates from the X_2 direction anti-clockwise by angle ϑ. Any stress component σ_{22}^* or σ_{12}^* is then given by the transformation relations:

$$\sigma_{22}^* = \sigma_{22}\cos^2\vartheta + \sigma_{33}\sin^2\vartheta + 2\sigma_{23}\sin\vartheta\cos\vartheta$$

or

$$\sigma_{23}^* = (\sigma_{33} - \sigma_{22})\sin\vartheta\cos\vartheta + \sigma_{23}(\cos^2\vartheta - \sin^2\vartheta)$$

that can be written for σ_{22n}^*, σ_{22m}^*, $\sigma_{22n}'^*$, $\sigma_{22m}'^*$, σ_{23n}^*, $\sigma_{23n}'^*$, σ_{23m}^* and $\sigma_{23m}'^*$.

The structural parameters η_{22n}^*, η_{22m}^* or η_{23n}^*, η_{23m}^* must be identical for any value of ϑ, i.e.:

$$\sigma_{22n}^* - \sigma_{22m}^* + \frac{\sigma_{22n}'^*}{\eta_{22n}^*} - \frac{\sigma_{22m}'^*}{\eta_{22m}^*} = 0$$

$$\sigma_{23n}^* - \sigma_{23m}^* + \frac{\sigma_{23n}'^*}{\eta_{23n}^*} - \frac{\sigma_{23m}'^*}{\eta_{23m}^*} = 0$$

If we use the transformation relations in the above two formulae and have in mind that these equations with fixed structural parameters must be valid for all values of ϑ, it turns out:

$$\eta_{22n}^* = \eta_{23n}^*, \quad \eta_{22m}^* = \eta_{23m}^*$$

and hence eqs. (253).

The procedure to show

$$\chi_{22n}^* = \chi_{23n}^*, \quad \chi_{22m}^* = \chi_{23m}^*$$

would be quite the same.

The validity of the relations

$$\eta_{12n} = \eta_{13n}, \ \eta_{12m} = \eta_{13m} \quad \text{and} \quad \chi_{12n} = \chi_{13n}, \ \chi_{12m} = \chi_{13m}$$

follows from the assumed transverse isotropy immediately.

REFERENCES

1/ Axelrad, D.R., "Micromechanics of Solids", PWN-Elsevier Publ. Co., Amsterdam /1976/.

2/ Bažant, Z.P. and Kim. S.S., "Plastic - Fracturing Theory of Concrete". J. Eng. Mech. Division 105, 407-428 /1979/.

3/ Beran, M.J., "Statistical Theory of Heterogeneous Media", p. 243, In: "Mechanics of Composite Materi als". ed. by F.W. Wendt. H. Liebowitz and N. Perrone, Pergamon Press /1970/.

4/ Bolshanina, M.A. and Panin, V.E., "The Stored Energy of Deformation" /in Russian/, In: Issledovanie po fizike tverdovo tela, Izd. AN SSSR. 193-233 /1957/. p.229.

5/ Bowden, Z.P. and Tabor, D., "The Influence of Size and Surface Structure on the Strength of Solids", In: The friction and lubrication of solids, Part II, Oxford at the Clarendon Press /1964/.

6/ Bruhns, O.T. and Müller, R., "Some Remarks on the Application of a Two-Surface Model in Plasticity", Acta Mechanica 53, 81-100 /1984/.

7/ Christensen, R.M., "Mechanics of Composite Materials", John Wiley & Sons, New York /1979/.

8/ Coleman, B.D. and Gurtin, M.E., "Thermodynamics with Internal Variables", The Journal of Chemical Physics 47. 597 /1967/.

9/ Dougill, J.W., "On Stable Progressively Fracturing Solids", ZAMP 27. 423-438 /1976/.

10/ Drucker. D.C., Prager W. and Greenberg, N.J., "Extended Limit Design Theorems for Continuous Media". Q. Appl. Math. 9, 381-389 /1952/.

11/ Edvards, R.H., "Stress Concentration around Spheroidal Inclusions and Cavities", Trans. ASME. J. Appl. Mech. 18, 1 /1951/.

12/ Eringen, A.C., "Nonlinear Theory of Continuous Media", Mc Graw - Hill, New York /1962/.

13/ Hashin. Z., "The Elastic Moduli of Heterogeneous Materials" Trans. ASME, J. Appl. Mech. 29, 143-150 /1962/.

14/ Hashin, Z., "Theory of Composite Materials", p. 201, In: "Mechanics of Composite Materials", ed. by F.W. Wendt, H. Liebowitz and N. Perrone, Pergamon Press, Oxford /1970/

15/ Hencky, H., "Zur Theorie plastischer Deformationen und der hierdurch im Material hervorgerufenen Nachspannungen", ZAMM 4, 323 /1924/.

16/ Hill. R., "The Mathematical Theory of Plasticity". Oxford at Clarendon Press, /1950/.

17/ Hill, R., "Elastic Properties of Reinforced Solids: Some Theoretical Principles". J.Mech.Phys.Solids 11, 357-372 /1963/.

18/ Hill, R., "The Essential Structure of Constitutive Laws for Metal Composites and Polycrystals", J.Mech.Phys.Solids 15, 79-95 /1967/.

19/ Hill. R., "On Macroscopic Effects of Heterogeneity in Elasto-plastic Media at Finite Strain", Math.Proc.Camb.Phil.Soc. 95, 481-494 /1984/.

20/ Hlaváček, M.and Kafka, V. "Non-Local Criteria of Strength" /in Czech/, Research report, Ústav teoretické a aplikované mechaniky ČSAV, Praha, /1986/.

21/ Inoue, T. and Yamamoto, K., "Static and Cyclic Deformations of Quasihomogeneous Elastic-Plastic Material", Proc. XXIIIrd Japan Congress on Materials Research, The Society of Materials Science, Kyoto /1980/, pp. 12-17.

22/ Janson, J. and Hult. J., "Fracture Mechanics and Damage Mechanics - a Combined Approach", Journal de mécanique appliquée 1, 69-84 /1977/.

23/ Jirků, J., "The Influence of the Rate of Loading on the Strength of an Al-Si Alloy" /in Czech/, MSc - Thesis, Faculty of Nuclear and Physical Engineering, Technical University, Prague /1981/.

24/ Jírová, J. and Kafka. V., "Structural Mathematical Model of the Rheologic Deformation of a Two-Component Material", Mechanics of Composite Materials /Transl. of "Mekhanika kompositnykh materialov" 18, 779-783 /1982// 18. 515-519 /1983/.

25/ Kafka, V., "Microstresses in Elastic-Plastic Deformation of Quasihomogeneous Materials", Rozpravy ČSAV, Series ofTechnical Sciences. No.4, /1972/.

26/ Kafka, V., "Determination of Material Constants for Quasi-homogeneous Strain-Hardening Materials", Acta technica ČSAV 17, 173-208 /1972/.

27/ Kafka. V., "Theory of Slow Elastic-Plastic Deformation of Polycrystalline Metals with Microstresses as Latent Variables Descriptive of the State of the Material", Archives of Mechanics 24, 403-418 /1972/.

28/ Kafka, V., "Structural Theory of Elastic-Plastic Deformation of Unidirectional Fiber-Reinforced Materials", Acta technica ČSAV, "Part I: The General Approach" 18, 418-436 /1973/, "Part II: The Constants of the Composite" 19, 87-95 /1974/.

29/ Kafka, V., "Zur Thermodynamik der plastischen Verformung", ZAMM 54, 649-657 /1974/.

30/ Kafka, V., "On Elastic-Plastic Deformation of Composite Materials with Anisotropic Structure", Acta technica ČSAV 21, 685-706 /1976/.

31/ Kafka, V., "On the Strength of Composite Materials in Elastic-Plastic Deformation", In: Proc. 13th Yugoslav Congress of Rational and Applied Mechanics, C4-4, 1-21 /1976/.

32/ Kafka, V., "On the Limit Analysis of Heterogeneous Materials", Acta technica ČSAV, "Part I: Theory" 22, 314-321 /1977/, "Part II: Examples of Application" 22, 706-717 /1977/.

33/ Kafka, V., "Strain-Hardening and Stored Energy", Acta technica ČSAV 24, 199-216 /1979/.

34/ Kafka, V. and Štětina, K., "Theory of Rheological Deformation of Quasi-Isotropic Eutectic Composites", Acta technica ČSAV 25, 205-224 /1980/.

35/ Kafka, V., "On the Hill's Fundamental Equation for Quasi-homogeneous Materials", ZAMM 63, 145-149 /1983/.

36/ Kafka, V., "Microstress Redistribution in Fibre-Reinforced Composites due to Time-Dependent Processes", In: Proc. Intern. Conf. "Mechanical Behaviour of Materials - IV" eds. J. Carlsson and N.G. Ohlson, Pergamon Press, Oxford /1983/, pp. 457-463.

37/ Kafka, V., and Jírová, J., "A Structural Mathematical Model for the Viscoelastic Anisotropic Behaviour of Trabecular Bone", Biorheology 20, 795-805 /1983/.

38/ Kafka, V., "Foundations of the Theoretical Microrheology of Heterogeneous Materials", /in Czech/, Academia, Prague, /1984/.

39/ Kafka, V., "Quantum Theory of Strength. Part I: Fundamentals", Acta technica ČSAV, 30, 137-146 /1985/.

40/ Kafka, V. and Němec, J., "The Reality of Fracture and the Model of Quantum Theory of Strength" /in Czech/, In: Vybrané práce Ústavu teoretické a aplikované mechaniky ČSAV z období 1981 až 1985, 67-74, Academia, Praha /1985/.

41/ Kafka, V, "Quantum Theory of Strength", ZAMM, 66, T 146-147 /1986/.

42/ Kratochvíl, J. and Dillon, O.W., "Thermodynamics of Elastic-Plastic Materials as a Theory with Internal State Variables". J. of Appl. Physics 40, 3207-3218 /1969/.

43/ Kruml, F., "Príspevok k relácii dotvarovania v tlaku a ťahu", In: Proc. Celostátní konference o betonu, Mariánské Lázně /1978/ pp. 97-115.

44/ Kunin, I.A., "Elastic Media with Microstructure I. One--Dimensional Models", Springer-Verlag, Berlin /1982/.

45/ Kunin, I.A., "Elastic Media with Microstructure II. Three-Dimensional Models", Springer-Verlag, Berlin /1983/.

46/ Lin. T.H., Salinas, D and Ito, Y.M., "Effects of Hydrostatic Stress on the Yielding of Cold Rolled Metals and Fiber-Reinforced Composites", J.Comp.Mat. 6, 409-413 /1972/.

47/ Lusche, M., "Beitrag zum Bruchmechanismus von auf Druck beanspruchtem Normal- und Leicht- Beton mit geschlossenem Gefüge", Beton-Verlag GmbH, Düsseldorf /1972/.

48/ Majumdar, S. and McLaughlin, R.V.Jr., "Application of Limit Analysis to Composite Materials and Structures", Trans. ASME, J.Appl.Mech. 41, 995-1000 /1974/.

49/ Majumdar, S. and McLaughlin, R.V.Jr., "Effects of Phase Geometry and Volume Fraction on the Plane Stress Limit Analysis of a Unidirectional Fiber-Reinforced Composite", Int.J. Solids and Struct. 11, 777-791 /1975/.

50/ "Mechanics of Composite Materials", ed. by G.P. Sendeckyj /vol.2 of "Composite Materials", ed. by L.J. Broutman and R.H. Krock/, Academic Press, New York /1974/.

51/ Muschelishvili, N.J., "Some Fundamental Problems of the Mathematical Theory of Elasticity" /In Russian/, Izd. AN SSSR, Moskva, /1954/.

52/ Naumenko. V.P. and Mitczenko, O.V., "Brittle Fracture of a Plate with a Hole in Compression" /in Russian/, Problemy procznosti, 7, 12-20 /1985/.

53/ Němec, J., "Rigidity and Strength of Steel Parts", Academia, Praha /1966/.

54/ Novotný, V. and Vích, M., "Einfluss der Oberflächenbeschädigung von Flachglass auf seine Festigkeit", In: Wissenschaftliche Beiträge der Friedrich Schiller - Universität Jena, Physikalische Chemie der Oberfläche, 144-160, /1984/.

55/ Peter, A. and Fehervary, A., "Evaluation of Acoustic Emission from Pressure Vessels with Planar Flaws", Theor. and Appl. Fracture Mechanics 5, 17-22 /1986/.

56/ Phillips, A., Tang, J.L. and Riccinti, M., "Some New Observations of Yield-Surfaces", Acta mech. 20, 23-39 /1974/.

57/ Phillips, A., "New Developments Concerning the Two-Surfaces Theory of Plasticity and Viscoplasticity", Trans.4th Int.Conf.Struct.Mech.React.Technol., San Francisco, Calif., 1977, Vol.1, L 1/5, pp.1-10, Amsterdam /1977/.

58/ Phillips, A. and Das, P.K., "Yield Surfaces and Loading Surfaces of Aluminum and Brass: An Experimental Investigation at Room and Elevated Temperatures", Int.J. of Plasticity 1, 89-109 /1985/.

59/ Pisarenko, G.S., Amelianowicz, K.K., Goralik, E.T., Brezneva, S.I. and Moskalenko, A.M., "The Dependence of Strength in Compression and Bending of Inorganic Glass on the Produced Thickness of the Plates" /in Russian/. Problemy procznosti, 4, 18-21, /1985/.

60/ Prager, W., "On Elastic, Perfectly Locking Materials", In: Proc. 11th Conf. Appl.Mechanics, München 1964, Springer Verlag, Berlin /1966/, pp. 538-544.

61/ Prewo, K.M. and Kreider, K.G., "High Strength Boron and Borsic Fiber Reinforced Aluminium Composites", J. of Composite Materials 6, 338 /1972/.

62/ Puch, V.P., "Strength and Rupture of Glass" /in Russian/, Nauka, Leningrad /1973/.

63/ Reiner, M., "Rheology", pp. 344-550, In: "Handbuch der Physik", Band VI., Springer-Verlag /1958/.

64/ Shah, S.P. and Slate, F.O., "Internal Microcracking, Mortar-Aggregate Bond and the Stress-Strain Curve of Concrete", In: Proc.Intern.Conf. of the Structure of Concrete and its Behaviour under Load, Cement and Concrete Assn., London, Sept. /1965/.

65/ Spooner, D.C. and Dougill, J.W., "A Quantitative Assessment of Damage Sustained in Concrete During Compressive Loading". Magazine of Concrete Research 27. 151-160 /1975/.

66/ "Steel Drawn Wire for Prestressing Reinforcement" /in Czech/, Czechoslovak Standard ČSN 426441.

67/ Stroeven, P., "Some Aspects of Micromechanics of Concrete", Stevin Laboratory, Technological University of Delft /1980/.

68/ Szczepiński, W., "On the Effect of Plastic Deformation on Yield Condition", Arch.Mech.Stos. 15, 275-296 /1963/.

69/ Titchener, A.L. and Bever, M.B., "The Stored Energy of Cold Work", Progress in Metal Physics 7, 247-338 /1958/,

p. 331 and p. 316.

70/ Thomas, T.C. et al., "Microcracking of Plain Concrete and the Shape of the Stress-Strain Curve", J. of the Amer. Concrete Institute, 209-224, Febr. /1963/.

71/ Valanis. K.C., "Proper Tensorial Formulation of the Internal Variable Theory. The Endochronic Time Spectrum", Archives of Mechanics 29, 173-185 /1977/.

72/ Valanis. K.C., "Some Fundamental Consequences of a New Intrinsic Time Measure", Archives of Mechanics 32, 171 /1980/.

73/ Van Fo Fy, G.A., "Theory of Reinforced Materials /in Russian/. Naukova Dumka, Kiev /1971/.

74/ Vasilev, D.M., "On the Microstresses Arising in Polycrystalline Specimens Subject to Plastic Deformation" /in Russian/, Zurn. Tekhn. Fiziki 28, 2527-2542 /1958/.

75/ Vasilev, D.M., "On the Microstresses Arising in Metals Subject to Plastic Deformation" /in Russian/, Fiz. Tverdogo Tela 1, 1736-1746 /1959/.

76/ Vasilev, D.M. and Kozevnikova, L.V., "On the Nature of Yield-Point Jog in Pure Iron and Carbon Steels" /in Russian/, Fiz. Tverdogo Tela 1, 1316-1319 /1959/.

77/ Vých, M., Novotný, V. and Kubišta, P., "Contribution à l´étude de la relation entre dimension du miroir de fracture et la résistance du verre", Verres réfract., 35, 838-844 /1981/.

78/ Zaitsev, J.W. and Wittmann, F.H., "Crack Propagation in a Two-Phase Material such as Concrete", Fracture 1977, vol. 3, ICF 4, Waterloo, Canada, June 19-24, /1977/.